AF352930

Effects of Land Use on the Ecohydrology of River Basin in Accordance with Climate Change

Effects of Land Use on the Ecohydrology of River Basin in Accordance with Climate Change

Editors

Wiktor Halecki
Dawid Bedla
Marek Ryczek
Artur Radecki-Pawlik

MDPI • Basel • Beijing • Wuhan • Barcelona • Belgrade • Manchester • Tokyo • Cluj • Tianjin

Editors
Wiktor Halecki
Polish Academy of Sciences
Poland

Dawid Bedla
University of Agriculture in
Krakow
Poland

Marek Ryczek
University of Agriculture in
Krakow
Poland

Artur Radecki-Pawlik
Cracow University of
Technology
Poland

Editorial Office
MDPI
St. Alban-Anlage 66
4052 Basel, Switzerland

This is a reprint of articles from the Special Issue published online in the open access journal *Land* (ISSN 2073-445X) (available at: https://www.mdpi.com/journal/land/special_issues/effects_land_use_ecohydrology_river_basin_accordance_climate_change).

For citation purposes, cite each article independently as indicated on the article page online and as indicated below:

LastName, A.A.; LastName, B.B.; LastName, C.C. Article Title. *Journal Name* **Year**, *Volume Number*, Page Range.

ISBN 978-3-0365-5645-1 (Hbk)
ISBN 978-3-0365-5646-8 (PDF)

Cover image courtesy of Dawid Bedla and Wiktor Halecki.

© 2022 by the authors. Articles in this book are Open Access and distributed under the Creative Commons Attribution (CC BY) license, which allows users to download, copy and build upon published articles, as long as the author and publisher are properly credited, which ensures maximum dissemination and a wider impact of our publications.

The book as a whole is distributed by MDPI under the terms and conditions of the Creative Commons license CC BY-NC-ND.

Contents

About the Editors

Wiktor Halecki

Wiktor Halecki (PhD) As a system data analyst, I specialize in computational ecology. Currently, I work as an assistant professor at the Institute of Nature Conservation Polish Academy of Sciences. Nature-based Solutions and blue–green infrastructure in urban areas are topics of my research. To conserve biodiversity, I am implementing ecosystem services, including the control of invasive species. Additionally, in my postdoctoral work, I contributed to an international project assessing how extreme weather events impact desertification using hydrological models. Among Wiktor Halecki's accomplishments is popularizing environmental protection knowledge. Over 370 articles of international scope were reviewed by me between 2017 and 2022. As a microblogger (@WiktorHalecki)/Twitter user, I present my research results in an accessible graphic form on various topics in biology and ecohydrology. A publication on forecasting climate change in drought-prone areas is in preparation.

Dawid Bedla

Dawid Bedla (PhD) In my research, I specialize in aquatic ecology, plant ecology and surface water pollution. I am currently working as an assistant professor at the University of Agriculture in Krakow. Water status assessment, blue–green infrastructure in urban areas and ecological indicators are the main topics of my research interests. I participated in an international project devoted to the implementation of bioeconomy, which was implemented from the budget of the "Horizon 2020" program supported by the European Union. I am the author of over 50 articles of international scope and a reviewer of 10 international publications in the years 2019–2022.

Marek Ryczek

Marek Ryczek (PhD) Marek Ryczek is an Associate Professor in Environmental Engineering, Mining and Power Engineering. He works at the University of Agriculture in Krakow (Faculty of Environmental Engineering and Surveying, Department of Land Reclamation and Environmental Development). His research is on soil water erosion, land reclamation, soil-science, and soil and water protection.

Artur Radecki-Pawlik

Artur Radecki-Pawlik (Prof.) Artur Radecki-Pawlik is a Professor of Civil Engineering (PhD, PE). He works at the Cracow University of Technology (Faculty of Civil Engineering). His research is on rivers and mountain streams, as well as low-head hydraulic structures, sediment transport, hydromorphology and fluvial geomorphology for engineers. He was previously employed at the Agricultural University of Krakow and by CBS and PBW HYDROPROJEKT Designing Office, Krakow.

 *land*

Editorial

Editorial: Effects of Land Use on the Ecohydrology of River Basins in Accordance with Climate Change

Wiktor Halecki [1,*], Dawid Bedla [2], Marek Ryczek [3] and Artur Radecki-Pawlik [4]

[1] Institute of Nature Conservation Polish Academy of Sciences, Mickiewicza 33, 31-120 Krakow, Poland
[2] Department of Ecology, Climatology and Air Protection, University of Agriculture in Krakow, 31-120 Krakow, Poland
[3] Department of Land Reclamation and Environmental Development, Faculty of Environmental Engineering and Surveying, University of Agriculture in Krakow, 31-120 Krakow, Poland
[4] Structural Mechanics and Material Mechanics, Faculty of Civil Engineering, Cracow University of Technology, 31-155 Krakow, Poland
* Correspondence: wiktor@mailmix.pl

Citation: Halecki, W.; Bedla, D.; Ryczek, M.; Radecki-Pawlik, A. Editorial: Effects of Land Use on the Ecohydrology of River Basins in Accordance with Climate Change. *Land* **2022**, *11*, 1662. https://doi.org/10.3390/land11101662

Received: 20 September 2022
Accepted: 22 September 2022
Published: 26 September 2022

Publisher's Note: MDPI stays neutral with regard to jurisdictional claims in published maps and institutional affiliations.

Copyright: © 2022 by the authors. Licensee MDPI, Basel, Switzerland. This article is an open access article distributed under the terms and conditions of the Creative Commons Attribution (CC BY) license (https://creativecommons.org/licenses/by/4.0/).

A total of nine original publications and one concept paper are included in this Special Issue on water management and land use (Appendix A). Research topics include desertification in the Mexico Valley, the evaluation of environmental conditions and chemicals in Lithuania's water supply, and land use changes in the Polish mountains. This Special Issue also discusses climatic and anthropogenic factors that influence the Yangtze River in China. Similarly, the article on the Yellow River catchment fits quite well within Land's scope.

Climate change has changed the precipitation phase in Europe. More rain and less snow are likely to be experienced in winter. Snow replenishes groundwater resources that feed aquatic ecosystems (rivers, reservoirs, lakes) and water-dependent ecosystems (wetlands, marshes, and peatlands). Consequently, water shortages could arise and worsen as a result of a lack of snow or increased variability and intensity of precipitation. A variety of economic sectors and agriculture systems need to be properly managed.

Waterways with altered morphology must be revitalized or rehabilitated in order to restore their natural corridors and connectivity with river valleys. Groundwater retention capacity and restoration are also affected by such activities. Biodiversity enhancement, with an ecohydrological system, warrants climate change adaptation.

As an alternative to renaturation measures, the development of small, flowing water reservoirs could be a viable solution for floodplain and river banks morphology (straightened, staggered, and connected with groundwater). It is important to focus on high-quality, reusable retention water. Ecohydrological studies improve the quality of water and the environment by integrating biological indicators and hydrological process. Nature-based Solutions to water management are environmentally friendly.

1. Water Management in Rural Areas

Water retention has been greatly reduced in recent decades as the hydrological system has been remodelled to favour runoff rather than storage [1]. In agricultural areas, overdeveloped drainage channels, as well as flood protection measures, such as building dikes and straightening rivers in response to heavy rain, increase water loss. The sustainable development of catchments depends on preserving, protecting, or restoring natural watercourses [2].

Water management must integrate surface and groundwater resources, water quality, and reuse with environmental concerns [3]. Extreme weather phenomena limit the resilience of hydrological systems. Water deficits can be reduced more frequently, and the quality of the water might be improved simultaneously [4]. By integrating preventive steps, the hydrological buffer of the landscape would be increased, and more water could be retained. A rural landscape is designed to store water and maintain groundwater resources [5]. Generally, local water authorities and decision-makers have minor influence on social capital. Rural areas should, then, include the following [6,7]:

- Promoting activities to increase humus in soils (which has been decreasing steadily in recent decades);
- A buffer zone and rewetting should be maintained along river banks with pastures in river valleys;
- Improvements in selected drainage areas through redevelopment,
- Replanting trees in river valleys and groundwater supply areas.

2. Blue–Green Infrastructure in an Urban Area

Spatial planning should integrate green areas and urban structures to adapt to climate change [8]. Water retention measures should be created as an element of city planning activities (natural, semi-natural, and artificial areas). Blue–green infrastructure is undoubtedly an integral part of the built fabric (facilities combining greenery and water). In addition to significant negative environmental impacts (such as the desertification of water-dependent ecosystems, a decrease in groundwater-fed peatlands, the disappearance of alder and riparian forests), it has major economic and social consequences from an ecohydrological perspective [9,10]. The water quality of transformed rivers with a simplified biological structure and reduced ability to self-purify will be adversely affected by pollutants and elevated temperatures in the cities.

3. Conclusions

As a result of climate change, surface water quality may deteriorate and groundwater may become scarce. There are a number of meteorological factors, such as torrential rainfall, that can increase surface runoff, causing an accumulation of pollutants. An ecohydrologist explores both ecological processes and hydrological cycles. Studies and research in this field may be helpful in assessing environmental hazards. Climate change adaptation and mitigation can also be achieved through changes in land use, especially in river valleys. Small farms with hydrological-based production may also suffer from reduced productivity due to uncontrolled groundwater demand. Hence, groundwater exploitation requires monitoring and protection, which should set new standards for water resources. Nature-based Solutions can be the subject of further research with an emphasis on the eco-hydrological approach, especially in cities.

Author Contributions: Conceptualization, W.H. and D.B.; validation, M.R.; formal analysis, A.R.-P.; writing—original draft preparation, W.H. and D.B. All authors have read and agreed to the published version of the manuscript.

Funding: This research received no external funding.

Data Availability Statement: Not applicable.

Conflicts of Interest: The authors declare no conflict of interest.

Appendix A

1. Ji, G.; Huang, J.; Guo, Y.; Yan, D. Quantitatively Calculating the Contribution of Vegetation Variation to Runoff in the Middle Reaches of Yellow River Using an Adjusted Budyko Formula. *Land* **2022**, *11*, 535. https://doi.org/10.3390/land11040535.
2. Zarzycki, J.; Korzeniak, J.; Perzanowska, J. Impact of Land Use Changes on the Diversity and Conservation Status of the Vegetation of Mountain Grasslands (Polish Carpathians). *Land* **2022**, *11*, 252. https://doi.org/10.3390/land11020252.
3. Borek, Ł.; Kowalik, T. Hydromorphological Inventory and Evaluation of the Upland Stream: Case Study of a Small Ungauged Catchment inWestern Carpathians, Poland. *Land* **2022**, *11*, 141. https://doi.org/10.3390/land11010141.
4. Česoniene, L.; ŠileikienėD.; Dapkienė, M. Influence of Anthropogenic Load in River Basins on RiverWater Status: A Case Study in Lithuania. *Land* **2021**, *10*, 1312. https://doi.org/10.3390/land10121312.

5.   Oleszczuk, R.; Zając, E.; Urbański, J.; Jadczyszyn, J. Rate of Fen-Peat Soil Subsidence Near Drainage Ditches (Central Poland). *Land* **2021**, *10*, 1287. https://doi.org/10.3390/land10121287.
6.   Borek, Ł.; Bogdał, A.; Kowalik, T. Use of Pedotransfer Functions in the Rosetta Model to Determine Saturated Hydraulic Conductivity (Ks) of Arable Soils: A Case Study. *Land* **2021**, *10*, 959. https://doi.org/10.3390/land10090959.
7.   Kruk, E.; Fudała, W. Concept of Soil Moisture Ratio for Determining the Spatial Distribution of Soil Moisture Using Physiographic Parameters of a Basin and Artificial Neural Networks (ANNs). *Land* **2021**, *10*, 766. https://doi.org/10.3390/land10070766.
8.   Ji, G.; Song, H.;Wei, H.; Wu, L. Attribution Analysis of Climate and Anthropic Factors on Runoff and Vegetation Changes in the Source Area of the Yangtze River from 1982 to 2016. *Land* **2021**, *10*, 612. https://doi.org/10.3390/land10060612.
9.   Fiedler, M. The Effects of Land Use on Concentrations of Nutrients and Selected Metals in Bottom Sediments and the Risk Assessment for Rivers of theWarta River Catchment, Poland. *Land* **2021**, *10*, 589. https://doi.org/10.3390/land10060589.
10.   Montero-Rosado, C.; Ojeda-Trejo, E.; Espinosa-Hernández, V.; Fernández-Reynoso, D.; Caballero Deloya, M.; Benedicto Valdés, G.S. Water Diversion in the Valley of Mexico Basin: An Environmental Transformation That Caused the Desiccation of Lake Texcoco. *Land* **2022**, *11*, 542. https://doi.org/10.3390/land11040542.

References

1.   Yang, H.; Piao, S.; Huntingford, C.; Ciais, P.; Li, Y.; Wang, T.; Peng, S.; Yang, Y.; Yang, D.; Chang, J. Changing the retention properties of catchments and their influence on runoff under climate change. *Environ. Res. Lett.* **2018**, *13*, 094019. [CrossRef]
2.   Bedla, D.; Halecki, W.; Król, K. Hydromorphological method and GIS tools with a web application to assess a semi-natural urbanised river. *J. Environ. Eng. Landsc. Manag.* **2021**, *29*, 21–32. [CrossRef]
3.   Battemarco, B.P.; Tardin-Coelho, R.; Veról, A.P.; de Sousa, M.M.; da Fontoura, C.V.T.; Figueiredo-Cunha, J.; Barbedo, J.M.R.; Miguez, M.G. Water dynamics and blue-green infrastructure (BGI): Towards risk management and strategic spatial planning guidelines. *J. Clean. Prod.* **2022**, *333*, 129993. [CrossRef]
4.   Donati, G.F.; Bolliger, J.; Psomas, A.; Maurer, M.; Bach, P.M. Reconciling cities with nature: Identifying local Blue-Green Infrastructure interventions for regional biodiversity enhancement. *J. Environ. Manag.* **2022**, *316*, 115254. [CrossRef]
5.   Hamel, P.; Tan, L. Blue–Green Infrastructure for Flood and Water Quality Management in Southeast Asia: Evidence and Knowledge Gaps. *Environ. Manag.* **2021**, *69*, 699–718. [CrossRef]
6.   Wilbers, G.-J.; de Bruin, K.; Seifert-Dähnn, I.; Lekkerkerk, W.; Li, H.; Ballinas, M.B.-P. Investing in Urban Blue–Green Infrastructure—Assessing the Costs and Benefits of Stormwater Management in a Peri-Urban Catchment in Oslo, Norway. *Sustainability* **2022**, *14*, 1934. [CrossRef]
7.   O'Donnell, E.; Netusil, N.; Chan, F.; Dolman, N.; Gosling, S. International Perceptions of Urban Blue-Green Infrastructure: A Comparison across Four Cities. *Water* **2021**, *13*, 544. [CrossRef]
8.   Dai, X.; Wang, L.; Tao, M.; Huang, C.; Sun, J.; Wang, S. Assessing the ecological balance between supply and demand of blue-green infrastructure. *J. Environ. Manag.* **2021**, *288*, 112454. [CrossRef] [PubMed]
9.   Flores, C.C.; Vikolainen, V.; Crompvoets, J. Governance assessment of a blue-green infrastructure project in a small size city in Belgium. The potential of Herentals for a leapfrog to water sensitive. *Cities* **2021**, *117*, 103331. [CrossRef]
10.   Hannah, D.M.; Wood, P.J.; Sadler, J. Ecohydrology and hydroecology: A new paradigm?? *Hydrol. Process.* **2004**, *18*, 3439–3445. [CrossRef]

 *land*

Article

The Effects of Land Use on Concentrations of Nutrients and Selected Metals in Bottom Sediments and the Risk Assessment for Rivers of the Warta River Catchment, Poland

Michał Fiedler

Department of Land Improvement, Environmental Development and Spatial Management, Faculty of Environmental Engineering and Mechanical Engineering, Poznan University of Life Sciences, 60-628 Poznań, Poland; michal.fiedler@up.poznan.pl

Abstract: Changes in the environment, aiming at agricultural intensification, progressive urbanisation and other forms of anthropopression, may cause an increase in soil erosion and a resulting increase in the pollution inflow to surface water. At the same time, this results in increased nutrient pollution of bottom sediments. In this study, the concentrations of total nitrogen (TN), total phosphorus (TP), total organic carbon (TOC), calcium (Ca), iron (Fe) and potassium (K) were analysed using bottom sediment samples collected at 39 sites located along the entire length of the Warta River and its tributaries. Agricultural use of land adjacent to rivers was found to significantly degrade sediment quality, while anthropogenic land use (as defined by Corine Land Cover classification—CLC), unlike previous studies, reduces the pollution loads in the bottom sediments. Forest use also contributes to the reduction of the pollution load in sediments. It was found that the significance of the relationship between pollutant concentrations and land use depends on the length of the river–land interface. According to the analyses, the level of correlation between the analysed constituents depends on the use of land adjacent to rivers. The impact of agricultural land use has the strongest effect in the 1 km zone and 5 km in the case of anthropogenic land use. The results showed that the variability of total phosphorus TP concentrations is strongly correlated with the variability of iron concentrations. SPI values indicate that the risk to sediment quality is low due to TOC and Fe concentrations. In contrast, the risk of sediment pollution by TN and TP shows greater differentiation. Although the risk is negligible for 40% of the samples, at the same time, for 33% of the samples, a very high risk of pollution with both TN and TP was found.

Keywords: sediment; nutrient element; risk; land cover; Warta River

Citation: Fiedler, M. The Effects of Land Use on Concentrations of Nutrients and Selected Metals in Bottom Sediments and the Risk Assessment for Rivers of the Warta River Catchment, Poland. *Land* **2021**, *10*, 589. https://doi.org/10.3390/land10060589

Academic Editors: Wiktor Halecki, Dawid Bedla, Marek Ryczek and Artur Radecki-Pawlik

Received: 16 May 2021
Accepted: 1 June 2021
Published: 2 June 2021

Publisher's Note: MDPI stays neutral with regard to jurisdictional claims in published maps and institutional affiliations.

Copyright: © 2021 by the author. Licensee MDPI, Basel, Switzerland. This article is an open access article distributed under the terms and conditions of the Creative Commons Attribution (CC BY) license (https://creativecommons.org/licenses/by/4.0/).

1. Introduction

River sediments are an important part of the cycling of materials in the aquatic ecosystem. The understanding of the pollution processes of river bottom sediments is of great importance because it can be an indicator of the ecological health of waters [1]. Sediments can accumulate pollutants and act as a buffer to absorb and release pollutants into the aquatic environment [2]. Loads of constituents such as carbon, nitrogen and phosphorus in bottom sediments and their interrelationships enable identification of the sources of organic matter and its transformations. The ratios of these constituents are affected by several environmental factors such as climate [3,4], terrestrial inflows, morphometry and use of adjacent areas [5] or the mineralogical composition of sediments [6]. Biochemical and biological transformations taking place in sediments and at the water–sediment interface [7,8], as well as the hydrodynamic processes of sediment transport in riverbeds [9], are also of great importance for concentrations and mutual proportions of constituents that can be found in bottom sediments.

The land-use pattern of the area located in the catchment is one of the more important factors affecting the concentrations of nutrients, as well as other pollutants, in river bottom sediments [10]. Intense agricultural production can be a significant source of nutrient inputs to surface waters, both as diffuse and point sources of pollution [11,12]. This impact varies over time and depends, among other things, on changes in agricultural production technology [13]. Szatten and Habel [14] showed that the decrease of intensive agricultural areas results in the reduction of sediments and nutrients loads flowing into the reservoir. Additionally, urbanised areas were indicated as a source of surface water pollution [15,16] and, consequently, also bottom sediment pollution. However, as indicated by Khatri and Tyagi [15], the anthropogenic factors affecting water pollution in rural areas differ from those in urban areas. The chemical composition of bottom sediments of the water bodies located in urban and industrial areas shows anthropogenic enrichment with organic matter, phosphorus, calcium and trace elements [17]. An increase in impermeable surfaces in these areas increases surface runoff and erosion processes, causing an increased inflow of sediments to rivers [18]. In turn, afforested areas enable a reduction in the inflow of pollutants to watercourses and, consequently, reduce their deposition in bottom sediments. Riparian forest buffer systems are an effective tool for controlling the number of sediments and related pollutants carried in surface runoff [19]. Forest ecosystems also reduce the leaching of soluble nutrients into stream water [20]. In this regard, there is a frequent emphasis on the role of riparian vegetation that can act as a buffer to intercept suspended sediments and pollutants [21]. Forested riparian zones can reduce loads of suspended sediment in the runoff by 90% [22]. Grass riparian filter strips show a similar reduction of sediments in runoff as observed in forested zones, while the reduction of nitrogen and phosphorus is near 50% [23,24]. However, using only the primary type of land use may not be sufficient to explain the influx of pollutants to the aquatic environment fully. Liu et al. [6] showed that more fragmented forms of land use would give higher pollutant loads even if they are of the same primary type of land use.

The assessment of a qualitative status of bottom sediments requires the adoption of certain standards defining the concentrations of chemicals, the exceeding of which causes the sediments to be considered polluted. This is mainly due to the simplicity and ease of their application in practice. These standards, depending on a local controller, may address a variety of substances, including organic carbon, nutrients, heavy metals, organic compounds and others, and have different thresholds [25,26].

The chemicals that make up bottom sediments are interrelated. According to previous studies, the relationships between TN, TP and TOC concentrations are the result of biochemical and geochemical processes occurring in the environment and enable the identification of the sources of pollution [27,28]. Carbon–nitrogen ratio is one of the main variables to determine the source of organic matter in rivers [29]. Yun and An [30] indicated that the N:P ratio could be used in diagnosing the ecological health of a stream. Additionally, the relationships between TP concentrations and Fe, Ca and TOC concentrations are indicated, but the links between these elements can be different for each river [31].

The researches on the quality of riverine bottom sediments in lowland areas of Poland conducted in recent years has focused mainly on the analysis of concentrations and hazards caused by heavy metals [32–34], while the risks of nutrients or organic matter have not been analysed in detail.

This study aims to analyse the spatial variability of TN, TP, TOC, Ca, Fe and K concentrations and their interrelationships in bottom sediments of the Warta River and its tributaries. The analysis also considered the effects of land cover structure and P, Ca and Fe concentrations in soils, found in areas adjacent to sediment sampling sites, on sediment composition for different lengths of the contact zone between the river and surrounding areas. This enables the assessment of sediment quality and the identification of factors affecting the concentrations of nutrients in sediments.

2. Materials and Methods

2.1. Site Description

The catchment area of the Warta River is 54,519 km². In terms of length, 808 km, it is the third-longest river in Poland (Figure 1). The tributaries of the Warta River, which were also analysed in the presented study, are the Prosna River; the Bobrowski Canal; the Mosiński Canal; the Wełna, Obra and Noteć rivers as well as the additional tributaries of the Noteć River: the Gwda and Drawa rivers (Figure 1). The source of the Warta River is 380 m above sea level, and its mouth to the Oder is 12 m above sea level. The catchment area of the Warta River varies in terrain. The upper-southern part of the catchment is located in the Kraków-Częstochowa Upland, passing through the Central Polish lowlands to the northern lake districts. The tributaries of the Warta River range from 11 km (the Bobrowski Canal) to 391 km (the Noteć River). Their areas range from 28 km² in the case of the Bobrowski Canal to 17,319 km² in the case of the Noteć River (Table 1).

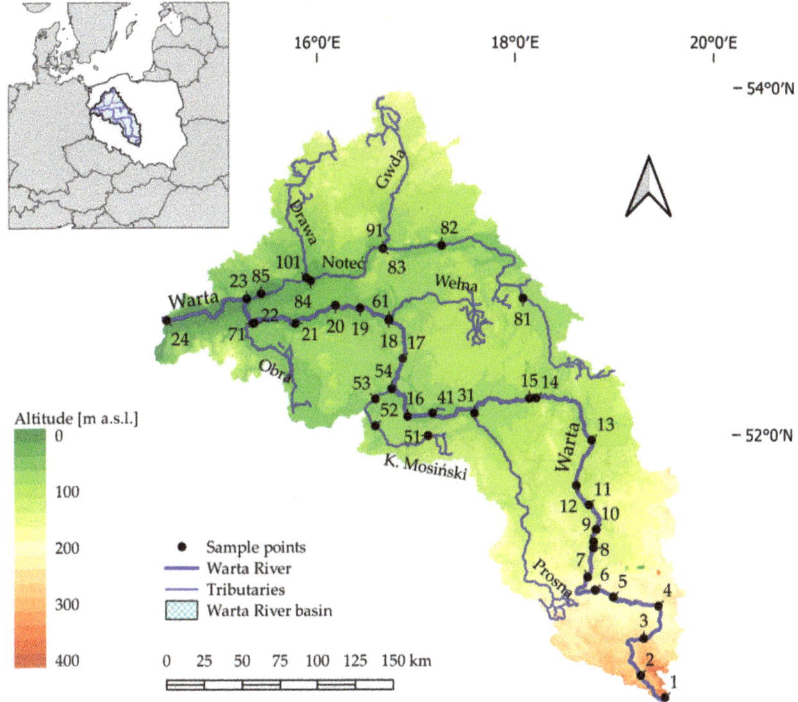

Figure 1. Study site location.

Almost along the entire length of the Warta river as well as its tributaries drains a post-glacial landscape formed during the retreat of the Vistulian ice-sheet. The layer of glacial sediments with a thickness of more than 200 m consists of tills, glacial sands and gravels and outwash sands and gravels [35]. In the Warta River catchment area, sandy soils make up 45% of the land area, clay soils 41% and organic soils and alluvial soils 14% of the land area [36]. According to the European Soil Database v2.0 m, the main soils are Podzol—32%, Luvisols—30%, Fluvisols—12% and Arenosols—12%.

According to the CLC classification, in the whole catchment area of the Warta River, agricultural land use makes up 60.2% of the catchment area, forest and semi-natural land use make up 32.5%, anthropogenic land use makes up 5.6%, surface water makes up 1.4% and the rest is covered by wetlands and peat moors (Table 1, Figure 2).

Table 1. Analysed rivers' descriptions.

River	Length (km)	Basin Area (km²)	Corine Land Cover (%)				
			1	2	3	4	5
Warta	808	54,519	5.6	60.2	32.5	0.2	1.4
Prosna *	227	4920	5.9	72.0	21.8	0.1	0.2
Kanał Bobrowski *	11	28	4.3	61.0	34.6	0	0
Kanał Mosiński *	142	2503	5.0	77.8	15.9	0.3	1.1
Wełna *	118	2620	3.6	72.0	22.5	0.1	1.8
Obra *	183	2759	3.8	50.7	43.7	0.2	1.6
Noteć *	391	17,319	3.1	51.0	43.1	0.3	2.4
Gwda **	145	4963	3.4	40.7	53.1	0.2	2.6
Drawa **	192	3290	2.0	32.2	61.8	0.1	3.9

Crine Land Cover: 1—Artificial surfaces, 2—Agricultural areas, 3—Forest and semi-natural areas, 4—Wetlands, 5—Water bodies; *—primary Warta tributary, **—secondary Warta tributary (via Noteć).

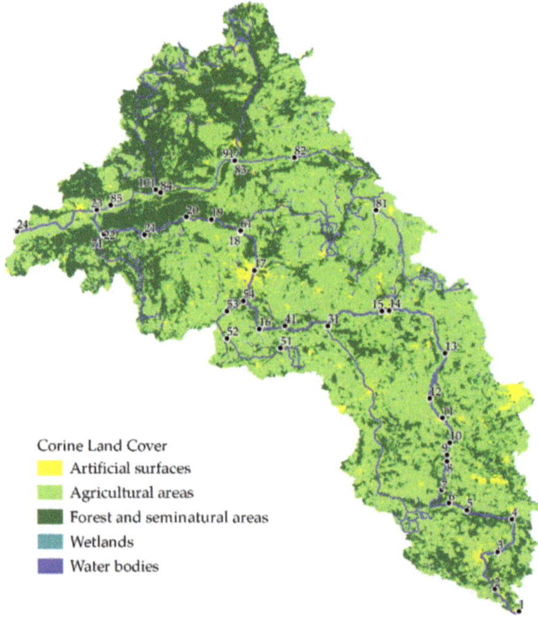

Figure 2. Land cover of the River Warta basin according to Corine Land Cover.

The highest proportion of agricultural land has the catchment area of the Mosiński Canal—approximately 78%. This catchment also has the lowest proportion of forest land and semi-natural land in a total catchment area—less than 16%. A slightly lower, compared to the aforementioned catchment, proportion of agricultural land in the total catchment area, 72%, is found in the catchment areas of the Prosna and Wełna rivers. On the other hand, the catchment area with the highest proportion of forest land and semi-natural forest land is the Drawa River catchment located in the northern part of the Warta catchment, where this proportion is 61.8%. This catchment also has the highest proportion of the open water area—3.9% of the total catchment area. The proportion of wetlands in all catchments is very small, reaching a maximum of 0.3% for the Mosiński Canal catchment.

An important source of bottom sediment contamination is the material brought into rivers as a result of erosion processes. One of the indicators that identify areas sensitive to erosion is the Topographic Wetness Index TWI. The TWI takes into account the upslope area and its slope, allowing the calculation of the steady-state wetness and runoff across the analysed area. Calculated TWIs show values in the range of 0.9–14.4 and are shown in Figure 3.

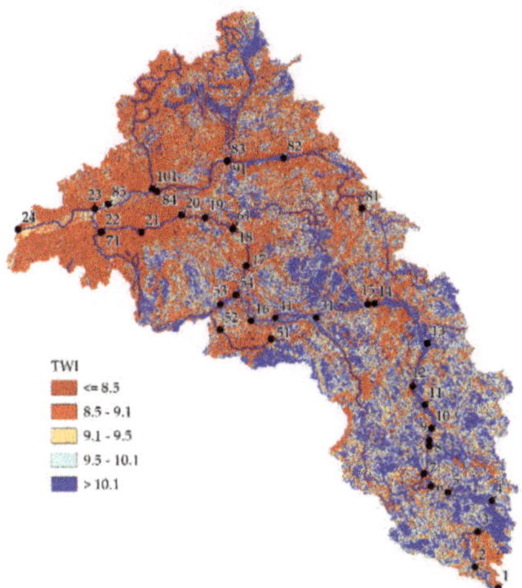

Figure 3. Distribution of Topographic Wetness Index TWI for the Warta River basin.

Geochemical conditions in the Warta River catchment, which describe the phosphorus, calcium and iron concentrations found in a 20 cm topsoil, are shown in Figure 4.

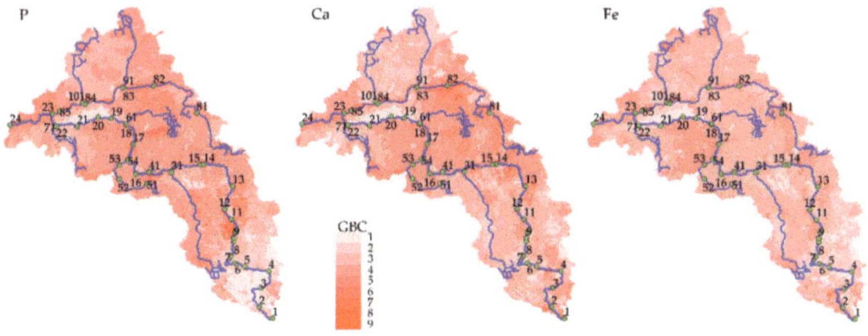

Figure 4. Classes of P, Ca and Fe content (GBC) in the upper layer of soils in the Warta River basin.

The Warta River catchment has the lowest annual precipitation in Poland [37]. Climatologically, according to the Köppen–Geiger's classification, the catchment area is located in the oceanic climate (Cfb) that has warm summers and milder winters [38,39]. The average annual air temperature for the WRC is approximately 8 °C, while the annual precipitation ranges from 520 mm in the north-eastern part of the catchment to 675 mm in the southern part [40]. The mean annual runoff is 3.86 $dm^3 \cdot s^{-1} \cdot km^{-2}$ for the Warta catchment and it shows considerable spatial variability [41]. The lowest runoff values, below 2 $dm^3 \cdot s^{-1} \cdot km^{-2}$, are observed in the upper catchment of the Noteć River [42].

2.2. Materials

The analyses used data concerning the measurements of N, P, K, Ca, Fe and TOC concentrations in bottom sediments of the Warta River and its tributaries, which were obtained as part of the State Monitoring of the Natural Environment programme. Bottom sediment samples were collected from 39 sites in 2016 (Figure 1). Sampling sites were located at the borders of the catchment area, at the mouths of tributaries and below cities where large industrial plants are located. For chemical analyses, a 5 cm top layer of sediments was collected from 4–5 sites over a 50 m distance for each location using a van Veen grab sampler. Samples for each site were mixed and rubbed through a nylon sieve with 2 mm mesh. Laboratory measurements were performed for Ca, K, Fe and P using inductively coupled plasma atomic emission spectrometry (ICP-AES) according to PN-EN 13657:2006, PN-EN ISO 11885:2009. Total organic carbon (TOC) was measured using coulometric titration according to CZ_SOP_D06_07_055 (CSN ISO 10694, CSN EN 13137). Kjeldahl nitrogen was determined using titration according to PN-EN 13342:2002.

The boundaries of the catchment area of the Warta River, its course and the course of its analysed tributaries were determined based on a Computer Map of the Hydrographic Division of Poland, which contains full hydrographic data of Poland in vector format. The land-use structure in the catchment area was made based on the CLC database provided by the Chief Inspectorate for Environmental Protection. The landform was described using a Digital Elevation Model (DEM) on a grid of at least 100 m, which was obtained from the Central Office of Geodesy and Cartography. The Ca, Fe and P concentrations found in the 20 cm topsoil were estimated using the Geochemical Atlas of Poland at a scale of 1:500,000, which was obtained from PGI—PIB (Polish Geological Institute—National Research Institute).

2.3. Methods

2.3.1. Overall Statistics

To evaluate the variability of N, P, K, Ca, Fe and TOC concentrations in bottom sediments of the analysed rivers, the mean values of the concentrations, the median and their maximum and minimum values were calculated. Concentration distributions were determined using the Shapiro–Wilk test. Outliers were determined using the Interquartile Range (IQR) test. The values that fall below or above 1.5-IQR were considered as outliers of the variables:

$$Up = Q_3 + 1.5 \cdot IQR \tag{1}$$

$$Low = Q_1 - 1.5 \cdot IQR \tag{2}$$

where Up—concentrations for values that fall above the 75th quartile (Q_3) and Low—concentrations for values that fall below the 25th quartile (Q_1). Correlations were determined using Spearman's rank coefficients.

All statistical calculations and visualisation of results were performed using R version 4.0.5.

2.3.2. Risk Assessment Method

An ecological risk assessment of nutrients in river bottom sediments was performed using a method based on the Single Pollution Index (SPI) [43].

The SPI_i is calculated based on the formula:

$$SPI_i = C_i / C_s \tag{3}$$

where C_i is a measured concentration of the evaluated i factor and C_s is a standard concentration of the evaluated i factor. Calculations were performed for TN, TP, TOC and Fe. Values of 550 mg·kg^{-1}, 600 mg·kg^{-1}, 1% and 2% were used as standard values for TN, TP, TOC and Fe, respectively, which are derived from safe nutrient concentration limits that can be found in the Sediment Quality Guidelines [25,43]. Based on the P_i, four hazard classes can be distinguished as shown in Table 2 [43].

Table 2. Single pollution index SPI_i classification.

Risk Level	TN	TP	TOC	Fe	Sediment Pollution
I	$P_{TN} < 1$	$P_{TP} < 0.5$	$P_{TOC} < 1$	$P_{Fe} < 0.5$	Clean
II	$1 \leq P_{TN} < 2$	$0.5 \leq P_{TP} < 1$	$1 \leq P_{TOC} < 5$	$0.5 \leq P_{Fe} < 1$	Slightly polluted
III	$2 \leq P_{TN} < 3$	$1 \leq P_{TP} < 1.5$	$5 \leq P_{TOC} < 10$	$1 \leq P_{Fe} < 2$	Moderately polluted
IV	$3 \leq P_{TN}$	$1.5 \leq P_{TP}$	$10 \leq P_{TOC}$	$2 \leq P_{Fe}$	Seriously polluted

C:P:N stoichiometry enables the description of potential sources of pollution as well as geochemical and biochemical processes occurring in the environment [31,44]. TOC/TN is one of more frequently used ratios, which identifies potential sources of organic matter [43]. TOC/TP reflects to some extent the rate of conversion of organic carbon and phosphorus compounds [45]. In contrast, TN/TP reflects the dynamics of accumulation, deposition and release of nitrogen and phosphorus in water [46,47]. TOC/TN > 10 indicates primarily a terrestrial source of organic matter. In contrast, TOC/TN < 10 indicates an aquatic source of organic matter. TOC/TN~10 indicates that the organic matter found in sediments is of both terrestrial and aquatic origin [48,49].

2.3.3. Analyses of Spatial Variability of Concentrations of Selected Elements Found in Bottom Sediments

The principal component analysis (PCA) and cluster analysis (CA) were used for the differentiation of groups of sampling sites exhibiting similar variability and similar sources. Due to concentrations of analysed substances in bottom sediments, the cluster analysis of sampling sites was performed using k-medoids [50,51]. K-medoids is a technique that is less sensitive to the effect of noise and outliers in analysed data, compared to the k-means. The number of cluster groups was found using the silhouette method [52].

Land cover was defined for river-adjacent zones of a width of 500 m, in % of the total area of the zone. The adopted zone lengths were 1, 2, 5 and 10 km from the sampling sites (Figure 5). In zones defined as above, GBI's content index for P, Ca and Fe was also calculated based on the formula:

$$GBI = \Sigma(GBC_i \cdot A_i)/\Sigma A_i \qquad (4)$$

where GBC_i means the topsoil content class for each element, while A_i is the area per class in an analysed zone.

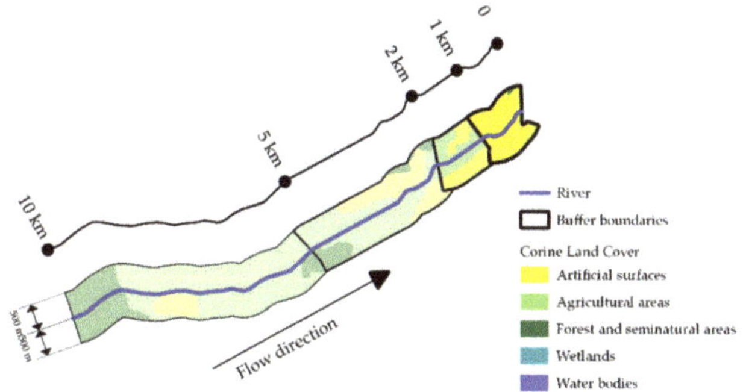

Figure 5. Scheme of buffer zones for land-cover analysis.

Spatial variability data concerning the parameters were determined using QGIS 3.18 whereas statistical calculations were performed using R 4.0.

3. Results and Discussion

3.1. General Characteristics of Concentration Variability

The results of measurements made in 2016 for TN, TP, Ca, Fe and TOC concentrations found in the bottom sediments of the Warta River and its tributaries are shown in Table A1. Concentrations of individual elements are within N: 71.5–16,240 mg-kg^{-3}, P: 48.8–8864 mg-kg^{-1}, Ca: 423–54,010 mg-kg^{-1}, Fe: 1602–68,280 mg-kg^{-1} and TOC: 0.20–21.50 % d.m.

The average concentrations of the analysed elements ranged from 13,182 mg-kg^{-1} for Fe through 9320 mg-kg^{-1} for Ca, 2169 mg-kg^{-1} for N to 1130 mg-kg^{-1} for P and 3.76% for TOC (Table 3). The median values are 8269 mg-kg^{-1}, 11,078 mg-kg^{-1}, 998 mg-kg^{-1}, 464 mg-kg^{-1} and 2.18% for Fe, Ca, N, P and TOC, respectively. The TOC values measured in 1998–2000 are similar to those measured in bottom sediments of the Oder, which is a receiving body for the water of the Warta River, when they ranged from 0.2% d.m. to 18.2% d.m., averaging 4.9% d.m [53]. Similar values for P, Ca, Fe and TOC concentrations were observed by House and Denison [31,54] in seven rivers in southern England. Their reported data concerning concentrations were 40–47,000 mg-kg^{-1} for Fe, 40–19,000 mg-kg^{-1} for Ca, 10–2000 mg-kg^{-1} for P and 0.6–19% d.m. for TOC. This may indicate similar sources of sediment pollution and similar land use of catchments.

Table 3. The descriptive statistics concentrations of chosen elements (mg·kg^{-1}) and TOC (% d.m.).

Site No.	TN	TP	TOC	Ca	Fe
min	71.5	48.8	0.20	423	1602
median	998	464	2.18	5985	8269
mean	2169	1130	3.76	9320	13,182
max	16,240	8864	21.50	54,010	68,280
range	16,169	8815.2	21.298	53,587	66,678
sd	3244	1619	4.70	11,078	12,930
skew	2.68	3.00	2.33	2.29	2.27
kurtosis	7.79	10.81	5.16	5.80	6.42
IQR	2685	1084.5	3.18	9934	13,821.5

The IQR values shown in Table 3 were 2685 mg-kg^{-1}, 1084 mg-kg^{-1}, 9934 mg-kg^{-1}, 13,821 mg-kg^{-1} and 3.18% for TN, TP, Ca, Fe and TOC, respectively. The IQR test showed several outliers involving N concentrations in the Warta River bottom sediments at site no. 4 and the Noteć River sediments at site no. 82. For phosphorus, concentration outliers were found in the Warta River sediments at sites no. 4, 19, 20 and in the Noteć River sediments, at site no. 82. For Ca, outliers were observed in the Mosiński Canal at sites no. 51 and 53. For Fe, outliers were found in the Noteć River at site no. 82. Outlier TOC values were observed in the Warta River sediments at site no. 4 and in the Noteć River bottom sediments at sites no. 82, 83 and 84. The outliers were most frequently observed at site no. 82 (Noteć River)—four times and three times higher/lower at sites no. 4 and 23 (Warta River).

The analysis of distributions of the analysed concentrations showed that none of the concentration distributions is a normal distribution (Figure 6). In contrast, for all $p > 0.05$ parameters, all distributions are similar to a log-normal distribution.

All analysed concentrations are strongly positively correlated with one another. Spearman's correlation coefficients range from 0.50 for the correlation between TOC and Ca to 0.93 for the correlation between TP and Fe. Almost all correlation coefficients are significant at the $p < 0.0001$ level (Table 4). Only the TOC–Ca correlation is significant at the $p < 0.01$ level, whereas the TOC–K correlation is significant at the $p < 0.001$ level. The strongest correlation is between Fe concentrations and TP, Ca and K concentrations, for which the correlation coefficients are 0.93, 0.87 and 0.86, respectively. In contrast, the correlations between TOC and Ca/K are the least correlated, but still significant at the $p < 0.01$ level.

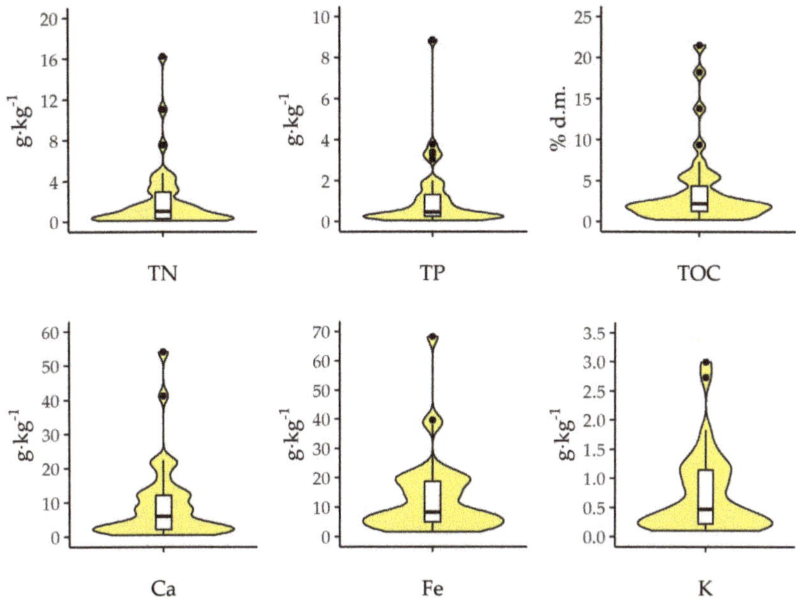

Figure 6. Variation of the concentrations of TN, TP, TOC, Ca, Fe and K in the bottom sediments of analysed rivers.

Table 4. Correlation coefficients between analysed elements.

	TN	TP	TOC	Ca	Fe
TP	0.79	-			
TOC	0.82	0.71	-		
Ca	0.74	0.76	0.50	-	
Fe	0.79	0.93	0.65	0.87	-
K	0.74	0.79	0.54	0.83	0.86

Inorganic constituents that can combine with or adsorb phosphates are one of the factors affecting TP concentrations in bottom sediments [55,56]. Immobilising components include metals, such as iron, which combine with phosphorus, forming crystalline metal phosphates (e.g., strengite or vivianite), or they can adsorb phosphorus on the oxide/hydroxide layer. In the case of the presence of large amounts of calcium, phosphate precipitation may occur in the form of hydroxylapatite (HA) or various calcium phosphates [55,57,58]. Furthermore, iron (III) plays an important role in the phosphate sorption by humic elements, forming iron (III)–humus complexes with phosphates [55,59].

3.2. C:N:P Stoichiometry

The calculated characteristic values of C:N, C:P and N:P ratios indicate their high variability as shown in Table 5 and Figure 7. C:N values range from 4.51 to nearly 112, with a median value of 22.98. The highest C:N values of 112, 81, 68 and 61 were observed at sites no. 11, 52, 3 and 21, respectively. A total of 33 sites had C:N > 10, two sites had ratios similar to 10, and only four sites had C:N < 10 (Figure 7). This indicates that the organic matter found in bottom sediments derives mainly from runoff from surrounding areas. Similarly, high C:N values were also obtained by Lu [60], indicating that exogenous sources of pollution are the cause. On the other hand, the lowest C:N values have sediments at sites located in the vicinity of the aforementioned site no. 52, i.e., sites no. 51 and 53–4.5 mg-kg^{-1} and 6.9 mg-kg^{-1}, respectively. The bottom sediments at these two sites have high nitrogen concentrations as well as very high P, Ca and Fe concentrations (Table A1).

Table 5. Descriptive statistics of C:N, C:P and N:P ratios.

Item	Min	Median	Mean	Max	1st Qu	3rd Qu	SD
C:N	4.51	22.98	30.1	111.85	12.90	39.65	22.91
C:P	6.83	37.04	58.3	501.27	20.55	63.70	82.68
N:P	0.28	1.47	2.32	15.37	1.18	2.40	2.61

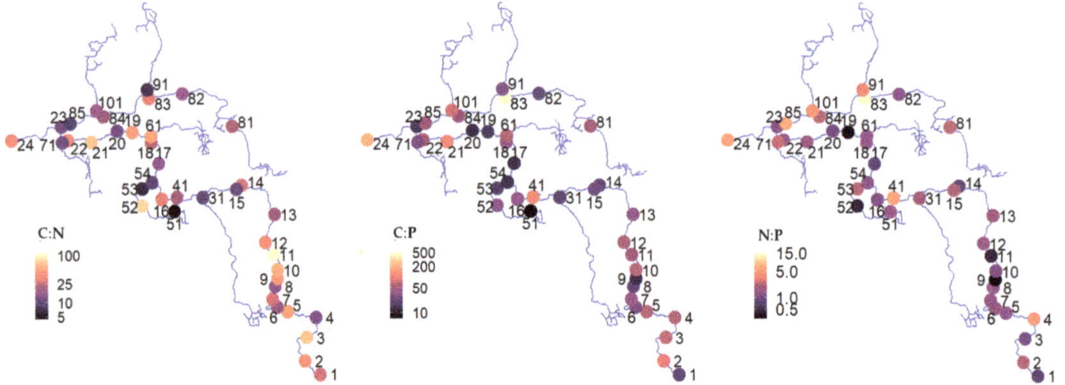

Figure 7. Spatial distribution of C:N, C:P and N:P ratios in the bottom sediments.

C:P ratio values in bottom sediments of rivers of the Warta River catchment ranged from 6.83 to 501.27, with a mean value of 58.3. The highest value of this parameter was obtained for site no. 83—the Noteć River, while the lowest for site no.—the Mosiński Canal.

N:P ratios in the bottom sediments of rivers of the Warta River catchment ranged from 0.28 to 15.37, with a mean value of 2.32. According to Chen et al. [45], low N:P values correspond to eutrophic and mesotrophic conditions with the predominant supply of nutrients derived from external sources. The nutrients from such sources are complex and have low N:P ratios. In contrast, high TN/TP values can be observed under oligotrophic conditions when natural nutrient supply sources have high N:P ratios. Knösche [55] reports that the TN:TP ratio in river sediments decreases significantly from low-density Paleopotamal sediments (median TN:TP = 23.7) to high-density Eupotamal sediments (median 8.5). He also observed a similar correlation for the TOC:TP ratio. At the same time, Ostendorp [61] indicated that N:P:OC ratios could be explained by the variability of organic content rather than by eutrophication processes in each case.

3.3. Single Pollution Index (SPI)

The pollution status of bottom sediments of the rivers of the Warta River catchment shows great spatial variation, as shown in Figure 8. SPI values for nitrogen range from 0.13 to 29.5. The largest proportion of sediment samples, 41%, can be classified as Class I (Table 6). Class II includes 13% of the samples. Sampling sites with a low risk of nitrogen pollution are located mainly in the upper part of the Warta River (sites no. 1–14, excluding sites no. 4 and 12) and in its lower reaches (sites no. 18–24, excluding site no. 20). On the other hand, Class III includes 13% of sediment samples and Class IV, with the highest pollution, includes as many as 33% of samples. These sites are located primarily in the middle course of the Warta River and in its tributaries.

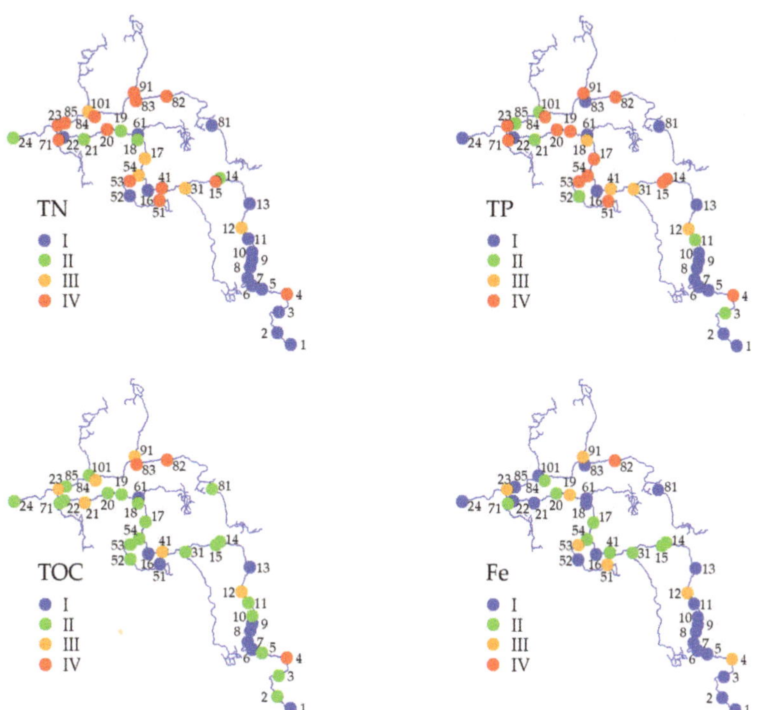

Figure 8. Spatial distribution of SPI for TN, TP, TOC and Fe.

Table 6. SPI classes distribution for Warta River basin in %.

Site No.	I	II	III	IV
TN	41	13	13	33
TP	39	15	10	36
TOC	23	54	15	8
Fe	56	23	18	3

SPI values for sediments polluted with phosphorus ranged from 0.08 to 14.8. Classification of sediment samples classifies 39% of samples into Class I, 15% into Class II, 10% into Class III and 36% into Class IV. These values are similar to SPI for TN. As with nitrogen, the sampling sites classified as Class I and II are located in the upper and lower reaches of the Warta River. The SPI values for TOC allow 23% of sediment samples to be classified as Class I and 54% as Class II. On the other hand, 15% and 8% of sediment samples should be classified into Class III and IV, respectively. This indicates a much lower risk of bottom sediment pollution by TOC than by TN or TP. The lowest pollution risk to bottom sediments was found for Fe. SPI values ranging from 0.08 to 3.4 enabled 56%, 23%, 18% and 3% of the samples to be classified as classes II–V, respectively.

The obtained SPI values indicate a lower pollution risk to bottom sediments of the Warta River than those obtained by He et al. [43] for Chinese rivers.

3.4. Spatial Variation of Pollutant Concentrations

The clustering analysis of TN, TP, TOC, Ca, Fe and K concentrations in bottom sediments using the Partitioning Around Medoids (PAM) identified three groups of bottom sediment sampling sites (Figure 9). A simultaneous principal component analysis (PCA) found that the first two principal components, for which the eigenvalues are greater than 1, explain approximately 87% of the total variance (Table 7).

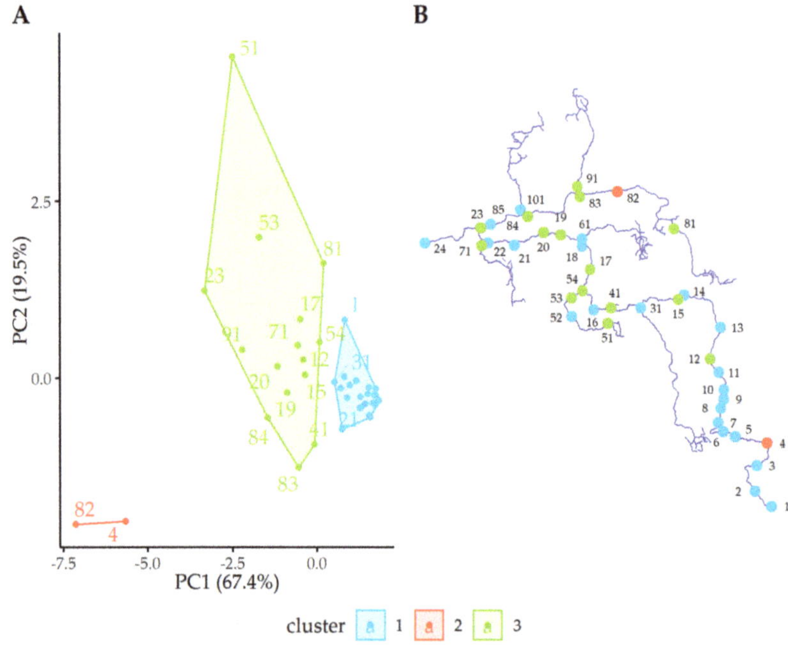

Figure 9. Partitioning clustering of sampling sites according to element concentrations in sediments (**A**) and location of points (**B**).

Table 7. Eigenvalues, percent of variance and cumulative percent of variance for the PCA.

Component.	Eigenvalue	Percent of Variance	Cumulative Percent of Variance
1	4.0463	67.44	67.44
2	1.1709	19.52	86.96
3	0.4460	7.43	94.39
4	0.2323	3.87	98.26
5	0.0799	1.33	99.59
6	0.0247	0.41	100.00

The land-use structure, as determined by the CLC classification, shows varying effects on the levels of constituents found in river sediments of the Warta River catchment. Anthropogenic lands (C_1) and forests and semi-natural ecosystems (C_3) show a negative correlation, while agricultural lands (C_2) and lands covered by water (C_5) show a positive correlation with concentrations of all elements; however, this relationship is not statistically significant in most cases. It should be noted that previous studies highlighted the negative impact of urbanised areas on the quality of bottom sediments in rivers. This was mainly due to the impact of direct discharge of wastewater into rivers or increasing surface runoff from areas with impervious surfaces [62,63]. However, the following phenomena that have been marked in recent years, such as the connection of most wastewater sources to wastewater treatment plants, the increase of areas that enable rainwater infiltration or, for example, measures to retain rainwater runoff from impervious surfaces, may reduce or neutralise the negative impact of urbanisation on the quality of surface water and bottom sediments. The negative impact of agricultural areas is mainly caused by an increased input of nutrients [12], inorganic suspended solids [64] and organic matter [65] from these areas to surface waters.

Factor loadings for the first two principal components are shown in Table 8. According to the classification presented by Liu et al. [66], factor loadings were classified as "strong," "average," and "weak" for values of >0.75, 0.75–0.50, and 0.50–0.30, respectively. PC1, which explains 67.4% of the total variance, has strong positive Fe, TN, TP and TOC loadings, while also showing average K and Ca loadings.

Table 8. Loading for PCA matrix.

	Components	
	1	**2**
TN	0.888 **	−0.295
TP	0.887 **	−0.172
TOC	0.798 **	−0.500 *
Ca	0.651 *	0.649 *
Fe	0.957 **	−0.045
K	0.704 *	0.617 *

**—strong effect; *—moderate effect.

Table 9 shows the Spearman's correlation coefficient values of TN/TP/K/Ca/Fe/TOC concentrations in bottom sediments with land cover structure in the zones of a width of 1000 m, adjacent to the river on a length of 1, 2, 5 and 10 km, and the calculated Ca, Fe and P content index of the topsoil.

The significance of the land-use impact depends not only on the land-cover structure itself, but also on the length of the contact zone of land adjacent to the watercourse. A significant ($p < 0.05$) effect of agricultural land (C_2) on the concentration levels of all analysed elements was observed for the 1 km buffer area. For longer buffer lengths, excluding K and TOC concentrations in the 2 km buffer area, that effect was no longer statistically significant. In contrast, the proportion of anthropogenic land (C_1) is significantly linked to P, K, Fe and TOC concentrations in the 5 km buffer area. Additionally, the proportion of the area covered by water (C_5) shows the strongest relationships with P, K, Ca and TOC concentrations in the 5 km buffer area. At the same time, the proportion of the land covered by water is the only one that is significantly correlated with phosphorus concentrations for all buffer lengths. Except for the effects of forest land proportion (C_3) on Fe concentrations in the 10 km buffer area, there was no significant impact of that land-use category on concentrations of analysed substances found in bottom sediments.

The results indicate a significant positive relationship between the geochemical background level of phosphorus in the 2 km buffer area and N, P, Ca, Fe and TOC concentrations in the bottom sediments analysed. In contrast, the geochemical background levels of Ca and Fe do not show such a relationship.

3.5. The Relationships between Water Quality Parameters and Land Use

The analysis of land use as identified based on the CLC data in individual buffers enabled four land-cover groups to be distinguished. The clustering results and the distribution of each group are shown in Figure 10. The first group includes areas dominated by anthropogenic lands, with a low proportion of forests. The second group includes buffer areas that are predominantly forested. The third group includes buffer areas dominated by agricultural lands. The fourth group consists of buffer areas where the proportion of lands covered by water is at least 10%, regardless of other land-use categories.

Table 9. Correlation coefficient matrix showing element–land cover and element–geochemical background relationships in sediments.

Buffer Length (km)	Parameter	N	P	TOC	Ca	Fe	K
1	C_1	−0.23	−0.28	−0.34 *	−0.28	−0.24	−0.30
	C_2	0.36 *	0.33 *	0.40 *	0.37 *	0.39 *	0.41 **
	C_3	−0.29	−0.14	−0.17	−0.31	−0.30	−0.31
	C_4	-	-	-	-	-	-
	C_5	0.09	0.35 *	0.19	0.24	0.13	0.19
2	C_1	−0.22	−0.26	−0.35 *	−0.27	−0.24	−0.31
	C_2	0.29	0.28	0.36 *	0.29	0.29	0.35 *
	C_3	−0.24	−0.11	−0.15	−0.22	−0.24	−0.27
	C_4	-	-	-	-	-	-
	C_5	0.12	0.34 *	0.21	0.27	0.17	0.24
5	C_1	−0.27	−0.32 *	−0.37 *	−0.26	−0.34 *	−0.32 *
	C_2	0.10	0.10	0.14	0.07	0.14	0.13
	C_3	−0.18	−0.03	−0.09	−0.13	−0.18	−0.18
	C_4	-	-	-	-	-	-
	C_5	0.20	0.42 **	0.32 *	0.34 *	0.25	0.32 *
10	C_1	0.00	−0.31	−0.19	−0.12	−0.07	−0.12
	C_2	0.14	0.22	0.21	0.17	0.24	0.19
	C_3	−0.30	−0.15	−0.20	−0.22	−0.33 *	−0.28
	C_4	0.04	0.25	0.20	−0.12	0.04	−0.04
	C_5	0.10	0.34 *	0.21	0.21	0.14	0.17
2	Ca_geo	0.17	0.22	0.18	0.17	0.05	0.15
	Fe_geo	0.24	0.16	0.24	0.08	0.19	0.07
	P_geo	0.36 *	0.35 *	0.45 **	0.32 *	0.41 **	0.29

*—significant at $p < 0.05$; **—significant at $p < 0.01$; Land Cover: C1—Artificial, C2—Agricultural, C3—Forests and seminatural, C4—Wetlands, C5—Water.

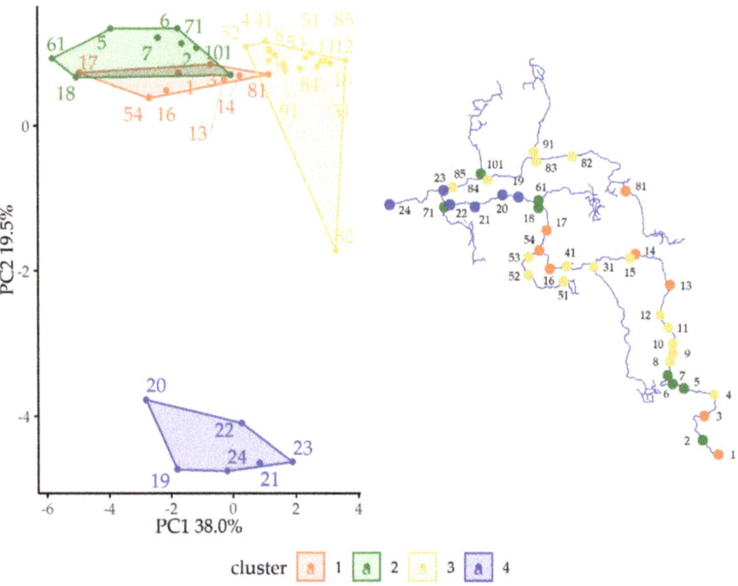

Figure 10. Classification of sampling points according to land use.

The Spearman's correlation coefficients between parameters describing constituent concentrations in bottom sediments for defined land-use groups are shown in Table A2. In group 1, where anthropogenic land use is the main factor, the strongest relationship is between TP and Fe, TN and TOC concentrations. However, the relationships between the elements show a higher correlation in the case of buffers that were identified based on a significant proportion of forest land. The highest correlation is between TP and TN, TN and TOC and TP and Fe. Unlike cluster 1, there is also a strong correlation between TOC and Fe in cluster 2.

The highest levels of correlation between bottom sediment constituents occur when the river is adjacent to agricultural land (cluster 3). In this case, all correlation coefficients are statistically significant, at least at the $p < 0.05$ level. The highest correlation in cluster 3 is between Ca and K, TP and Fe and Fe and K. The lands included in cluster 4 have surface water within the buffer area, the proportion of which is between 10% and 20% (highest proportion) of the total buffer area. In the case of sediment samples collected at sites assigned to this cluster, the highest correlation occurred between Ca and K concentrations, where r = 1.0. Again, very high levels of correlation occurred between Fe and TP and Ca and K.

It should be noted that regardless of the influence of the environment, phosphorus concentrations show the highest correlation with iron concentrations and slightly lower correlation with calcium concentrations, which is due to the already mentioned mutual interactions between these constituents, leading to precipitation or release of phosphorus [67–69]. The strong correlations between TP and TN concentrations for all use categories of the river-adjacent zones, except for the relatively low correlation for agricultural use, may indicate similar sources of nutrients in these areas. In agricultural lands, however, sources of pollution are of different nature—diffuse or point. Nutrients may derive from both field crops and animal husbandry, and they have different nutrient ratios, as also indicated by Lee [70] and Jones [10]. At the same time, bottom sediments collected from sites included in cluster 3 have the highest average TN, TOC, Ca and Fe concentrations and also high TP concentrations (Table 10). In contrast, the concentration values for clusters 1 and 2 (i.e., river-adjacent zones with anthropogenic and forest land uses) are significantly lower. Average sediment constituent concentrations for cluster 3 compared to concentrations for clusters 1 and 2 are 4.3–5.8 times greater for nitrogen, 2.3–4.0 times greater for phosphorus, 3.4–4.2 times greater for TOC, 4.3–5.8 times greater for calcium and 1.9–2.8 times greater for iron. The intermediate values of constituent concentrations were found in the sediments for cluster 3 determined on the basis of the proportion of surface water in the river-adjacent zone. As this proportion is between 10–20%, the remaining area of the land may have very different uses and, consequently, average out the effects of different land uses on the bottom sediment composition.

Table 10. Mean concentrations of elements (mg·kg^{-1}) and TOC (%) in sediments samples for land-use clusters.

Cluster	TN	TP	TOC	Ca	Fe	K
1	634	622	1.35	8375	9153	1025
2	839	370	1.67	3976	6159	387
3	3651	1475	5.76	13,015	17,339	855
4	1792	1840	4.06	7236	16,138	941

4. Conclusions

Based on the results of analyses concerning spatial variability of N, P, TOC, Ca, Fe and K concentrations in the bottom sediments of the Wart River and its tributaries, the following conclusions can be made:

- The dominance of agricultural activities in the zone of land adjacent to the river was found to be a significant factor increasing the pollutant content found in the bottom sediments of the analysed rivers. On the other hand, artificial and wooded areas reduce the concentrations of analysed substances in bottom sediments. The significance of this impact also depends on the length of the contact zone between the river and the land adjacent to it.

- The results of C:N:P ratios indicate that the organic matter found in sediments has mainly a terrestrial source. These results also indicate that there is a threat of eutrophication to a substantial number of the rivers.
- Based on the analysis of SPI values, it can be concluded that the risk resulting from TOC and Fe is relatively low or absent in almost 80% of the sampling sites, and it is medium to high in the remaining sampling sites. On the other hand, SPI values for TN and TP indicate that although the risk is absent for 40% of the samples, at the same time, more than a third of the samples are severely polluted.
- Future analyses will require consideration of the temporal variability of pollutant concentrations. Numerous studies indicate that sediment parameters vary not only over annual periods but also over multi-year periods [32]. This is due to the influence of river processes on sediment composition, in addition to the inflow of substances from adjacent areas, due to different water flow rates and the resulting migration of pollutants.

Funding: The publication was co-financed within the framework of the Ministry of Science and Higher Education programme as "Regional Initiative Excellence" in years 2019–2022, Project No. 005/RID/2018/19.

Data Availability Statement: The data presented in this study are available on request from the corresponding author.

Conflicts of Interest: The author declares no conflict of interest.

Appendix A

Table A1. Concentrations of chosen elements (mg·kg^{-1}) and TOC (% d.m.) in the bottom sediments of the Warta river.

River	Site No.	TN	TP	TOC	Ca	Fe	K
	1	207	276	0.568	22,430	9460	170
	2	491	217	1.98	630	3200	129
	3	331	320	2.27	2232	6640	216
	4	16,240 *	3405 *	21.5 *	21,810	38,180	1384
	5	260	193	1.24	3398	4270	212
	6	354	220	0.643	1705	3608	100
	7	71.5	48.8	0.202	423	2518	332
	8	156	108	0.233	2010	3044	137
	9	82.7	294	0.375	3068	4877	237
	10	330	283	1.84	1457	4840	255
	11	211	464	2.36	2050	6874	227
Warta	12	1380	881	5.15	8420	20,300	1192
	13	127	71.9	0.26	760	2017	117
	14	701	927	1.93	4081	12,760	511
	15	3040	1419	3.49	8486	15,870	950
	16	219	186	0.709	808	2626	116
	17	1380	1756	2.18	11,470	17,550	1428
	18	998	683	2.53	3467	8269	440
	19	1100	3280 *	4.99	8288	21,160	879
	20	3070	3054 *	3.86	11,990	19,380	1119
	21	935	573	6.22	3285	6640	325
	22	367	251	1.43	3763	8368	470
	23	4630	3773 *	5.51	15,090	39,680 *	2731 *
	24	650	108	2.38	999	1602	122

Table A1. *Cont.*

River	Site No.	TN	TP	TOC	Ca	Fe	K
Prosna	31	1560	837	1.66	3188	13,800	726
K. B. [1]	41	3490	631	7.34	5985	14,310	320
Kanał Mosiński	51	1860	1229	0.839	54,010 *	22,070	2994 *
	52	188	427	1.52	2637	7734	276
	53	3950	1638	2.71	41,250 *	20,050	1197
	54	1560	1213	1.6	11,590	14,000	890
Wełna	61	169	126	0.818	1915	1964	211
Obra	71	2920	1145	3.42	12,160	18,610	1169
Noteć	81	544	228	1.25	13,630	8174	1828
	82	11,040 *	8864 *	18.2 *	21,200	68,280 *	1208
	83	4230	275.3	13.8 *	8257	7041	556.1
	84	4790	2000	9.37 *	12,990	18,750	874.6
	85	1970	332	1.74	6388	4937	416.8
Gwda	91	7550 *	1992	5.78	18,050	23,800	1584
Drawa	101	1450	324.2	2.55	8108	6835	505.7

[1] Kanał Bobrowski, * outliers.

Table A2. The correlation coefficients between the concentrations of elements in the bottom sediments for buffer zones with various land covers.

	TN	TP	TOC	Ca	Fe
		Anthropogenic land cover (cluster 1, $n = 8$)			
TP	0.86 **				
TOC	0.69	0.76 *			
Ca	0.36	0.40	0.05		
Fe	0.83 *	0.93 ***	0.52	0.62	
K	0.76*	0.62	0.52	0.60	0.69
		Forest land cover (cluster 2, $n = 8$)			
TP	0.95 ***				
TOC	0.93 ***	0.83 *			
Ca	0.79 *	0.79 *	0.86 **		
Fe	0.86 **	0.90 **	0.81 *	0.83 *	
K	0.57	0.52	0.69	0.74 *	0.69
		Agricultural land cover (cluster 3, $n = 17$)			
TP	0.74 ***				
TOC	0.86 ***	0.60 *			
Ca	0.76 ***	0.80 ***	0.50 *		
Fe	0.75 ***	0.92 ***	0.62 **	0.88 ***	
K	0.74 ***	0.81 ***	0.49 *	0.93 ***	0.92 ***
		Water land cover (cluster 4, $n = 6$)			
TP	0.89 *				
TOC	0.60	0.60			
Ca	0.83 *	0.89 *	0.26		
Fe	0.77	0.94 **	0.31	0.94 **	
K	0.83 *	0.89 *	0.26	1.00 ***	0.94 **

*—$p < 0.05$, **—$p < 0.01$, ***—$p < 0.001$; n—number of samples.

References

1. Liao, J.; Chen, J.; Ru, X.; Chen, J.; Wu, H.; Wei, C. Heavy metals in river surface sediments affected with multiple pollution sources, South China: Distribution, enrichment and source apportionment. *J. Geochem. Explor.* **2017**, *176*, 9–19. [CrossRef]
2. Petticrew, E.L.; McConnachie, J.L. Sediment-water interactions. In *Sediment Dynamics and Pollutant Mobility in Rivers*; Westrich, B., Förstner, U., Eds.; Environmental Science and Engineering; Springer: Berlin/Heidelberg, Germany, 2007; pp. 217–268. ISBN 978-3-540-34782-8.

3. Ruttenberg, K.C.; Goñi, M.A. Phosphorus distribution, C:N:P ratios, and Δ13Coc in arctic, temperate, and tropical coastal sediments: Tools for characterizing bulk sedimentary organic matter. *Mar. Geol.* **1997**, *139*, 123–145. [CrossRef]

4. Owens, P.N. Sediment behaviour, functions and management in river basins. In *Sustainable Management of Sediment Resources*; Elsevier: Amsterdam, The Netherlands, 2008; Volume 4, pp. 1–29. ISBN 978-0-444-51961-0.

5. Walling, D.E. Tracing suspended sediment sources in catchments and river systems. *Sci. Total Environ.* **2005**, *344*, 159–184. [CrossRef]

6. Liu, A.; Duodu, G.O.; Goonetilleke, A.; Ayoko, G.A. Influence of land use configurations on river sediment pollution. *Environ. Pollut.* **2017**, *229*, 639–646. [CrossRef]

7. Fillos, J.; Swanson, W.R. The release rate of nutrients from river and lake sediments. *J. Water Pollut. Control Fed.* **1975**, *47*, 1032–1042.

8. Qian, Y.; Liang, X.; Chen, Y.; Lou, L.; Cui, X.; Tang, J.; Li, P.; Cao, R. Significance of biological effects on phosphorus transformation processes at the water–sediment interface under different environmental conditions. *Ecol. Eng.* **2011**, *37*, 816–825. [CrossRef]

9. Symader, W.; Bierl, R.; Kurtenbach, A.; Krein, A. Transport indicators. In *Sediment Dynamics and Pollutant Mobility in Rivers*; Westrich, B., Förstner, U., Eds.; Environmental Science and Engineering; Springer: Berlin/Heidelberg, Germany, 2007; pp. 269–304. ISBN 978-3-540-34782-8.

10. Johnes, P.J. Evaluation and management of the impact of land use change on the nitrogen and phosphorus load delivered to surface waters: The export coefficient modelling approach. *J. Hydrol.* **1996**, *183*, 323–349. [CrossRef]

11. Jarvie, H.P.; Withers, P.J.A.; Bowes, M.J.; Palmer-Felgate, E.J.; Harper, D.M.; Wasiak, K.; Wasiak, P.; Hodgkinson, R.A.; Bates, A.; Stoate, C.; et al. Streamwater phosphorus and nitrogen across a gradient in rural–agricultural land use intensity. *Agric. Ecosyst. Environ.* **2010**, *135*, 238–252. [CrossRef]

12. Dauer, D.M.; Weisberg, S.B.; Ranasinghe, J.A. Relationships between benthic community condition, water quality, sediment quality, nutrient loads, and land use patterns in Chesapeake Bay. *Estuaries* **2000**, *23*, 80. [CrossRef]

13. Renwick, W.H.; Vanni, M.J.; Fisher, T.J.; Morris, E.L. Stream nitrogen, phosphorus, and sediment concentrations show contrasting long-term trends associated with agricultural change. *J. Environ. Qual.* **2018**, *47*, 1513–1521. [CrossRef] [PubMed]

14. Szatten, D.; Habel, M. Effects of land cover changes on sediment and nutrient balance in the catchment with cascade-dammed waters. *Remote Sens.* **2020**, *12*, 3414. [CrossRef]

15. Khatri, N.; Tyagi, S. Influences of natural and anthropogenic factors on surface and groundwater quality in rural and urban areas. *Front. Life Sci.* **2015**, *8*, 23–39. [CrossRef]

16. Peters, N.E.; Meybeck, M. Water quality degradation effects on freshwater availability: Impacts of human activities. *Water Int.* **2000**, *25*, 185–193. [CrossRef]

17. Machowski, R.; Rzetala, M.A.; Rzetala, M.; Solarski, M. Anthropogenic enrichment of the chemical composition of bottom sediments of water bodies in the neighborhood of a non-ferrous metal smelter (Silesian Upland, Southern Poland). *Sci. Rep.* **2019**, *9*, 14445. [CrossRef]

18. O'Driscoll, M.; Clinton, S.; Jefferson, A.; Manda, A.; McMillan, S. Urbanization effects on watershed hydrology and in-stream processes in the Southern United States. *Water* **2010**, *2*, 605–648. [CrossRef]

19. Lowrance, R.; Altier, L.S.; Newbold, J.D.; Schnabel, R.R.; Groffman, P.M.; Denver, J.M.; Correll, D.L.; Gilliam, J.W.; Robinson, J.L.; Brinsfield, R.B.; et al. Water quality functions of riparian forest buffers in Chesapeake Bay watersheds. *Environ. Manag.* **1997**, *21*, 687–712. [CrossRef]

20. Qualls, R.G.; Haines, B.L.; Swank, W.T.; Tyler, S.W. Retention of soluble organic nutrients by a forested ecosystem. *Biogeochemistry* **2002**, *61*, 135–171. [CrossRef]

21. Vigiak, O.; Malagó, A.; Bouraoui, F.; Grizzetti, B.; Weissteiner, C.J.; Pastori, M. Impact of current riparian land on sediment retention in the Danube River Basin. *Sustain. Water Qual. Ecol.* **2016**, *8*, 30–49. [CrossRef]

22. McKergow, L.A.; Weaver, D.M.; Prosser, I.P.; Grayson, R.B.; Reed, A.E.G. Before and after riparian management: Sediment and nutrient exports from a small agricultural catchment, Western Australia. *J. Hydrol.* **2003**, *270*, 253–272. [CrossRef]

23. Daniels, R.B.; Gilliam, J.W. Sediment and chemical load reduction by grass and riparian filters. *Soil Sci. Soc. Am. J.* **1996**, *60*, 246–251. [CrossRef]

24. Yuan, Y.; Bingner, R.L.; Locke, M.A. A review of effectiveness of vegetative buffers on sediment trapping in agricultural areas. *Ecohydrology* **2009**, *2*, 321–336. [CrossRef]

25. Persaud, D.; Jaagumagi, R.; Hayton, A. *Guidelines for the Protection and Management of Aquatic Sediment Quality in Ontario*; Queen's Printer for Ontario: Toronto, ON, Canada, 1993.

26. Burton, G.A. Sediment quality criteria in use around the world. *Limnology* **2002**, *3*, 65–76. [CrossRef]

27. Peñuelas, J.; Poulter, B.; Sardans, J.; Ciais, P.; van der Velde, M.; Bopp, L.; Boucher, O.; Godderis, Y.; Hinsinger, P.; Llusia, J.; et al. Human-induced nitrogen–phosphorus imbalances alter natural and managed ecosystems across the globe. *Nat. Commun.* **2013**, *4*, 2934. [CrossRef] [PubMed]

28. Zhang, W.; Jin, X.; Liu, D.; Tang, W.; Shan, B. Assessment of the sediment quality of freshwater ecosystems in Eastern China based on spatial and temporal variation of nutrients. *Environ. Sci. Pollut. Res.* **2017**, *24*, 19412–19421. [CrossRef]

29. Venkatesh, M. Appraisal of the carbon to nitrogen (C/N) ratio in the bed sediment of the Betwa River, Peninsular India. *Int. J. Sediment Res.* **2020**, *35*, 69–78. [CrossRef]

30. Yun, Y.-J.; An, K.-G. Roles of N:P ratios on trophic structures and ecological stream health in lotic ecosystems. *Water* **2016**, *8*, 22. [CrossRef]

31. House, W.A.; Denison, F.H. Total phosphorus content of river sediments in relationship to calcium, iron and organic matter concentrations. *Sci. Total Environ.* **2002**, *282–283*, 341–351. [CrossRef]

32. Jaskuła, J.; Sojka, M.; Fiedler, M.; Wróżyński, R. Analysis of spatial variability of river bottom sediment pollution with heavy metals and assessment of potential ecological hazard for the Warta River, Poland. *Minerals* **2021**, *11*, 327. [CrossRef]

33. Sojka, M.; Siepak, J.; Jaskuła, J.; Wicher-Dysarz, J. Heavy metal transport in a river-reservoir system: A case study from Central Poland. *Pol. J. Environ. Stud.* **2018**, *27*, 1725–1734. [CrossRef]

34. Boszke, L.; Sobczyński, T.; Glosińska, G.; Kowalski, J.; Siepak, J. Distribution of mercury and other heavy metals in bottom sediments of the Middle Odra River (Germany/Poland). *Pol. J. Environ. Stud.* **2004**, *13*, 495–502.

35. Galon, R. *Morphology of the Noteć–Warta (or Toruń–Eberswalde) Ice Marginal Streamway*; Geographical Studies; Wydawnictwa Geologiczne: Warszawa, Poland, 1961; pp. 1–129.

36. Ilnicki, P.; Górecki, K.; Melcer, B. *Eutrophication of Rivers in the Warta Catchment in 1992–2002*; Poznań University of Life Sciences: Poznań, Poland, 2008; ISBN 978-83-7160-531-4.

37. Ilnicki, P.; Farat, R.; Górecki, K.; Lewandowski, P. Long-term air temperature and precipitation variability in the Warta River catchment area. *J. Water Land Dev.* **2015**, *27*, 3–13. [CrossRef]

38. Kottek, M.; Grieser, J.; Beck, C.; Rudolf, B.; Rubel, F. World map of the Köppen-Geiger climate classification updated. *Meteorol. Z.* **2006**, *15*, 259–263. [CrossRef]

39. Rubel, F.; Brugger, K.; Haslinger, K.; Auer, I. The climate of the European Alps: Shift of very high resolution Köppen-Geiger climate zones 1800–2100. *Meteorol. Z.* **2017**, *26*, 115–125. [CrossRef]

40. Climate Maps of Poland. Available online: https://klimat.imgw.pl/pl/climate-maps/ (accessed on 22 April 2021).

41. Ilnicki, P.; Farat, R.; Górecki, K.; Lewandowski, P. Impact of climatic change on river discharge in the driest region of Poland. *Hydrol. Sci. J.* **2014**, *59*, 1117–1134. [CrossRef]

42. Pasławski, Z.; Koczorowska, J.; Olejnik, K. Average outflow in the Warta River catchment. *Water Manag.* **1972**, *6*, 214–218.

43. Ye, H.; Yang, H.; Han, N.; Huang, C.; Huang, T.; Li, G.; Yuan, X.; Wang, H. Risk assessment based on nitrogen and phosphorus forms in watershed sediments: A case study of the upper reaches of the Minjiang Watershed. *Sustainability* **2019**, *11*, 5565. [CrossRef]

44. Gu, Y.-G.; Ouyang, J.; Ning, J.-J.; Wang, Z.-H. Distribution and sources of organic carbon, nitrogen and their isotopes in surface sediments from the largest mariculture zone of the Eastern Guangdong Coast, South China. *Mar. Pollut. Bull.* **2017**, *120*, 286–291. [CrossRef] [PubMed]

45. Chen, M.; Zeng, G.; Zhang, J.; Xu, P.; Chen, A.; Lu, L. Global landscape of total organic carbon, nitrogen and phosphorus in lake water. *Sci. Rep.* **2015**, *5*, 15043. [CrossRef]

46. Chlot, S. Nitrogen and Phosphorus Interactions and Transformations in Cold-Climate Mine Water Recipients. Ph.D. Thesis, Department of Civil Environmental and Natural Resources Engineering, Luleå University of Technology, Luleå, Sweden, 2013.

47. Yang, Y.; Gao, B.; Hao, H.; Zhou, H.; Lu, J. Nitrogen and phosphorus in sediments in China: A national-scale assessment and review. *Sci. Total Environ.* **2017**, *576*, 840–849. [CrossRef]

48. Mathew, J.; Gopinath, A.; Martin, G.D. Spatial and temporal distribution in the biochemical composition of sedimentary organic matter in a tropical estuary along the West Coast of India. *SN Appl. Sci.* **2019**, *1*, 150. [CrossRef]

49. Stein, R. Accumulation of organic carbon in Baffin Bay and Labrador Sea sediments (ODP-Leg 105). In *Accumulation of Organic Carbon in Marine Sediments*; Lecture Notes in Earth Sciences; Springer: Berlin/Heidelberg, Germany, 1991; Volume 34, pp. 40–84. ISBN 978-3-540-53813-4.

50. Schubert, E.; Rousseeuw, P.J. Faster k-Medoids clustering: Improving the PAM, CLARA, and CLARANS algorithms. In *Similarity Search and Applications*; Amato, G., Gennaro, C., Oria, V., Radovanović, M., Eds.; Lecture Notes in Computer Science; Springer International Publishing: Cham, Switzerland, 2019; Volume 11807, pp. 171–187. [CrossRef]

51. Kassambara, A. *Practical Guide To Cluster Analysis in R. Unsupervised Machine Learning*. Available online: https://www.datanovia.com/en/product/practical-guide-to-cluster-analysis-in-r/ (accessed on 2 June 2021).

52. Batool, F.; Hennig, C. Clustering with the average silhouette width. *Comput. Stat. Data Anal.* **2021**, *158*, 107190. [CrossRef]

53. Niemirycz, E.; Gozdek, J.; Koszka-Maron, D. Variability of organic carbon in water and sediments of the Odra River and its tributaries. *Pol. J. Environ. Stud.* **2006**, *15*, 557–563.

54. House, W.A.; Denison, F.H. Nutrient dynamics in a lowland stream impacted by sewage effluent: Great Ouse, England. *Sci. Total Environ.* **1997**, *205*, 25–49. [CrossRef]

55. Knösche, R. Organic sediment nutrient concentrations and their relationship with the hydrological connectivity of floodplain waters (River Havel, NE Germany). *Hydrobiologia* **2006**, *560*, 63–76. [CrossRef]

56. Watson, S.J.; Cade-Menun, B.J.; Needoba, J.A.; Peterson, T.D. Phosphorus forms in sediments of a river-dominated estuary. *Front. Mar. Sci.* **2018**, *5*, 302. [CrossRef]

57. Richardson, C.J.; Reddy, K.R. Methods for soil phosphorus characterization and analysis of wetland soils. In *Methods in Biogeochemistry of Wetlands*; John Wiley & Sons, Ltd.: Hoboken, NJ, USA, 2013; pp. 603–638. ISBN 978-0-89118-961-9.

58. Stevenson, F.J.; Cole, M.A. *Cycles of Soil: Carbon, Nitrogen, Phosphorus, Sulfur, Micronutrients*, 2nd ed.; Wiley: New York, NY, USA, 1999; ISBN 978-0-471-32071-5.

59. Håkanson, L.; Jansson, M. *Principles of Lake Sedimentology*; Springer: Berlin, Germany; New York, NY, USA, 1983.
60. Lu, S.; Si, J.; Qi, Y.; Wang, Z.; Wu, X.; Hou, C. Distribution characteristics of TOC, TN and TP in the wetland sediments of Longbao Lake in the San-Jiang Head waters. *Acta Geophys.* **2016**, *64*, 2471–2486. [CrossRef]
61. Ostendorp, W. Sedimente und Sedimentbildung in Seeuferröhrichten des Bodensee-Untersees. *Limnologica* **1992**, *22*, 16–33.
62. Slaymaker, O. Land use effects on sediment yield and quality. In *Sediment/Freshwater Interaction*; Sly, P.G., Ed.; Springer: Dordrecht, The Netherlands, 1982; pp. 93–109. ISBN 978-94-009-8011-2.
63. Rimoldi, F.; Peluso, L.; Bulus Rossini, G.; Ronco, A.E.; Demetrio, P.M. Multidisciplinary approach to a study of water and Bottom sediment quality of streams associated with mixed land uses: Case study Del Gato Stream, La Plata (Argentina). *Ecol. Indic.* **2018**, *89*, 188–198. [CrossRef]
64. Crosbie, B.; Chow-Fraser, P. Percentage land use in the watershed determines the water and sediment quality of 22 marshes in the Great Lakes Basin. *Can. J. Fish. Aquat. Sci.* **1999**, *56*, 1781–1791. [CrossRef]
65. Marval, Š.; Hejduk, T.; Zajíček, A. An analysis of sediment quality from the perspective of land use in the catchment and pond management. *Soil Sediment Contam. Int. J.* **2020**, *29*, 397–420. [CrossRef]
66. Liu, C.-W.; Lin, K.-H.; Kuo, Y.-M. Application of factor analysis in the assessment of groundwater quality in a blackfoot disease area in Taiwan. *Sci. Total Environ.* **2003**, *313*, 77–89. [CrossRef]
67. Chen, M.; Ding, S.; Chen, X.; Sun, Q.; Fan, X.; Lin, J.; Ren, M.; Yang, L.; Zhang, C. Mechanisms driving phosphorus release during algal blooms based on hourly changes in iron and phosphorus concentrations in sediments. *Water Res.* **2018**, *133*, 153–164. [CrossRef] [PubMed]
68. Van der Grift, B.; Osté, L.; Schot, P.; Kratz, A.; van Popta, E.; Wassen, M.; Griffioen, J. Forms of phosphorus in suspended particulate matter in agriculture-dominated lowland catchments: Iron as phosphorus carrier. *Sci. Total Environ.* **2018**, *631–632*, 115–129. [CrossRef] [PubMed]
69. Smolders, A.J.P.; Lamers, L.P.M.; Lucassen, E.C.H.E.T.; Van Der Velde, G.; Roelofs, J.G.M. Internal eutrophication: How it works and what to do about it—A review. *Chem. Ecol.* **2006**, *22*, 93–111. [CrossRef]
70. Lee, J.Y.; Yang, J.S.; Kim, D.K.; Han, M.Y. Relationship between land use and water quality in a small watershed in South Korea. *Water Sci. Technol.* **2010**, *62*, 2607–2615. [CrossRef] [PubMed]

 *land*

Article

Hydromorphological Inventory and Evaluation of the Upland Stream: Case Study of a Small Ungauged Catchment in Western Carpathians, Poland

Łukasz Borek * and Tomasz Kowalik

Department of Land Reclamation and Environmental Development, University of Agriculture in Kraków, al. Mickiewicza 24–28, 30-059 Krakow, Poland; tomasz.kowalik@urk.edu.pl
* Correspondence: lukasz.borek@urk.edu.pl

Abstract: The hydromorphological conditions of watercourses depend on numerous natural and anthropogenic factors such as buffer zones or human infrastructure near their banks. We hypothesised that, even in a small stream, there can be substantial differences in the hydromorphological forms associated with naturalness and human impact. The paper aims at the field inventory and evaluation of the hydromorphological conditions of a small upland stream in the conditions of contemporary human activity, against the background of meteorological and hydrological conditions. The study concerned a left-bank tributary of the Stradomka River located in the Wiśnicz Foothills (Western Carpathians). The analyses were conducted with the use of the Polish method, the Hydromorphological Index for Rivers (HIR), which conforms to the EU Water Framework Directive (WFD). The hydromorphological condition and quality of habitats were evaluated based on the Hydromorphological Diversity Score (HDS) and Habitat Modification Score (HMS). The study shows that the largest changes in stream hydomorphology and habitat conditions took place in the downstream, urbanised stream catchment area with an intensive development of construction and technical infrastructure. The hydromorphological condition of the examined stream sections was evaluated as good or poor. The best hydromorphological conditions were found in the section located in the semi-natural area, and the worst in the urbanised area. As our research shows, the strong influence of human activity, including weather extremes, and the risks and hydrological hazards of the hydromorphological conditions of the small, ungauged catchment, highlight the necessity to search for other research methods to support the decision-making cycle in the transformation of riverbeds and catchments.

Keywords: hydromorphological diversity; highland watercourse; human activity; catchment management; weather extremes

Citation: Borek, Ł.; Kowalik, T. Hydromorphological Inventory and Evaluation of the Upland Stream: Case Study of a Small Ungauged Catchment in Western Carpathians, Poland. *Land* **2022**, *11*, 141. https://doi.org/10.3390/land11010141

Academic Editor: Ilan Stavi

Received: 21 December 2021
Accepted: 14 January 2022
Published: 17 January 2022

Publisher's Note: MDPI stays neutral with regard to jurisdictional claims in published maps and institutional affiliations.

Copyright: © 2022 by the authors. Licensee MDPI, Basel, Switzerland. This article is an open access article distributed under the terms and conditions of the Creative Commons Attribution (CC BY) license (https://creativecommons.org/licenses/by/4.0/).

1. Introduction

Upland and mountainous rivers, streams and brooks have large flow variations during a year, which are caused by higher precipitation than in the lowlands, as well as the varied topography—especially the gradient (slope) of the river which, for upland and mountain streams, generally ranges from 2% to 7% [1]. The density of the river network is greater than on the lowlands; hence, the dynamics related to water flow and changes in the river catchment are significant. The evaluation of the ecology conditions/potential of a WFD body of surface water, to date, has mainly focused on larger watercourses (classified as rivers) with a catchment area greater than 10 km^2, omitting their numerous tributaries (streams, brooks); the latter permanently or temporarily supply them with water, and are considered—often incorrectly—as artificial watercourses (e.g., ditches) formed as a result of land reclamation. Some of these small watercourses, often called 'wild streams', have never been examined in terms of their hydromorphology. Some of them are unnamed, not supervised and are not classified in terms of their natural or artificial origin. The beds of numerous, but mainly small mountain and upland streams are often dry, and are classified as 'temporary sporadic', while heavy rainfall causes streams and rivers to swell, sometimes leading to landslides. The Carpathian streams also experience high water erosion which carves the flysch bedrock [2]. An important part of understanding a stream's hydromorphology is to undestand the catchment's geological structure and topography [3,4]. The Carpathian region (including the Wiśnicz Foothills) is built by flysch formations, Cretaceous and Eocene shales, marls and sandstones. They are covered with several inches of loess layers from Quaternary sediments [5]. The stream catchment area can be divided into three sections: upper, middle and lower [6]; these three sections are clearly distinguishable in mountain streams, and are hard to distinguish in upland streams. The heads of mountain and upland streams are usually located on mountain sides covered with trees or other vegetation.

The role of such watercourses in mountainous and upland areas is the natural drainage of the catchment areas [7]. Far-reaching anthropogenic changes in recent years, such as the development/sealing of river catchment areas, amongst other changes, cause various problems. Often, such small watercourses/streams/brooks are significantly changed or eliminated, e.g., for land development. The anthropogenic modifications involving the transformation of open riverbeds, streams and brooks into concrete pipelines can increase the occurrence of inundations and so-called flash floods [8–10]. Studies show that changes in climate and the use of land and water must be considered together to fully understand watershed hydromorphology [11]. A great deal of attention is given to the pollution of surface waters, and not enough attention has been paid to the effects of transformations in the beds and micro-catchments of small upland and mountain streams that form an integral part of larger river catchment areas.

The term 'hydromorphology' relates to the hydrology and geomorphology of a small watercourse, and includes the analysis of the physical attributes of the watercourse bed, such as flow type, material of the bottom and bank slopes, channel modification, and its natural elements; it also includes the characteristics of the land-use within 5 m and 50 m of the bank-top [12,13]. Hydromorphology is one of the key elements of the ecological integrity of waters, according to the EU Water Framework Directive WFD [14]. Member States shall, in accordance with this directive and European Standards EN 14614 [15] and EN 15843 [16], standardise provisions regarding the hydromorphological evaluation of a watercourse, allowing a better understanding of its functioning, and, if necessary, correct regulation of the riverbed [6]. The basic term used in hydromorphology is catchment, i.e., the area limited by the watershed in which the water flows to one place (the main watercourse/recipient). This is a dynamic system that depends on natural factors. The catchment morphodynamics vary and depend on elements of the natural environment, such as geological structure, climate and weather conditions, landform, river gradient/slope, soil and flora [17]. Under unfavourable conditions (e.g., flash floods), a seemingly small stream can cause damage to both a catchment area and human property.

Hydromorphology still occupies a small space in the water environment management of Europe and Poland. With the constant pressure of climatic change and changes in politics, society and economy, the restoration of natural habitats in our water environment is often regarded as a non-priority, despite the fact that hydromorphology supports the biodiversity of our waters. Good hydromorphological condition is a basic element of ecosystem health and a foundation of many ecosystem services and benefits for society [18]. The hydromorphological evaluation of a watercourse can be used for planning the restoration of rivers and streams by means of their regulation or biotechnical development [6]. There is evidence that even a small hydromorphological interaction can have a deep impact on the functioning of an ecosystem [19]. There are numerous methods for the hydromorphological evaluation of watercourses [20–22]. The most commonly used methods for assessing the quality of physical river habitats are: the German LAWA-FS [23]; the British River Habitat Survey, RHS [6]; the Spanish assessment of bank habitats, QBR; and the Czech comprehensive morphological assessment, HEM [24]. These methods are based on field surveys and the characterization of physical attributes of riverbeds and flow regimes, and can be classified as river habitat surveys or physical habitat assessments [20]; despite the different assumptions, the methods lead to similar results and can be used in various countries, especially in Europe [24]. From the groups of methods used for field inventory of rivers for their restoration purposes, those recommended are: the Australian River Styles Framework [25]; the IHG method in Spain [26]; the MQI method in Italy [27]; and the Polish method of river hydromorphological quality assessment RHQ [28]. The index-based methods are being developed to allow the standardisation of methods and tools used for evaluation of hydromorphology and impact on the the ecological condition of watercourses/catchment areas [29]. In Poland, the Hydromorphological Index for Rivers (HIR) method, which conforms to the EU Water Framework Directive (WFD), has been used since 2017. The method provides the assessment of lowland, upland and mountain rivers and streams, and it can be used to evaluate natural and heavily modified watercourses, as well as artificial channels [30]. The HIR method is based on the British River Habitat Survey method, which is widely used for the hydromorphological assessment of waters around the world, as demonstrated by a number of citations in other research papers more than 2,000,000 times (on the Google Scholar website). The majority of the research on the hydromorphological assessment of watercourses has focused on large river systems. This work concentrates on the hydromorphological evaluation of a small upland stream, whose catchment area is being transformed as a result of high human activity. It is considered that, for the small ungauged aquatic ecosystem, it is advisable to know the hydromorphological conditions, which affect the flow of water and biodiversity.

We hypothesised that, even in a small stream, there is a significant variety of hydromorphological forms associated with naturalness and anthropopressure. The paper aims to provide a field inventory and evaluation of the hydromorphological conditions of a small ungauged stream catchment using the HIR method, including the meteorological and hydrological conditions.

2. Materials and Methods

The stream catchment area is located in southern Poland (Figure 1), in the Wiśnicz Foothills, part of the Western Carpathians. Empirical studies were conducted in the bed and valley of the 'unnamed stream', a left-bank tributary of the Stradomka river of the left-bank tributary of the Raba river (in the Upper Vistula Basin). The unnamed stream is a natural, periodic watercourse. In the currently valid typology of surface waters [31], it is classified as an upland carbonate stream with fine-grained substrate on loess and loess-like rocks (type code: 6). The studies were divided into 3 stages:

(1)   meteorological and hydrological conditions;
(2)   physiographic parameterisation of the stream catchment area;
(3)   hydromorphological characteristics of the watercourse based on HIR.

In terms of climate, according to the Köppen-Geiger classification [32], the examined catchment is located in a warm, humid continental climate (Dfb). The datasets from 2 meteorological stations from a multi-year period (2001–2020) were used in the analysis. The monthly precipitation totals (from the Łapanów station; GPS coordinates: 49°51′44″ N, 20°16′32″ E) and monthly average temperatures (from the Łazy station; GPS coordinates: 49°57′54″ N, 20°29′43″ E), provided by the monitoring network system and carried out by the Institute of Meteorology and Water Management–National Research Institute (IMGW-BIP), was used in the study. Both stations are located in the basin of the Upper Vistula within a distance of 20 km. The precipitation characteristics for monthly and annual periods were made on the basis of the relative precipitation index (RPI). The RPI classifies precipitation in terms of its excess or shortage [33]. The precipitation and thermal characteristics of the year 2020 were presented against the background of the long-term period 2001–2020. The RPI coefficient was calculated using the following formula:

$$RPI = \frac{P}{\overline{P}} \cdot 100 \ (\%) \tag{1}$$

where P—precipitation sum in the studied period (mm), and $\overline{P}$—average precipitation value in the studied long-term period (mm).

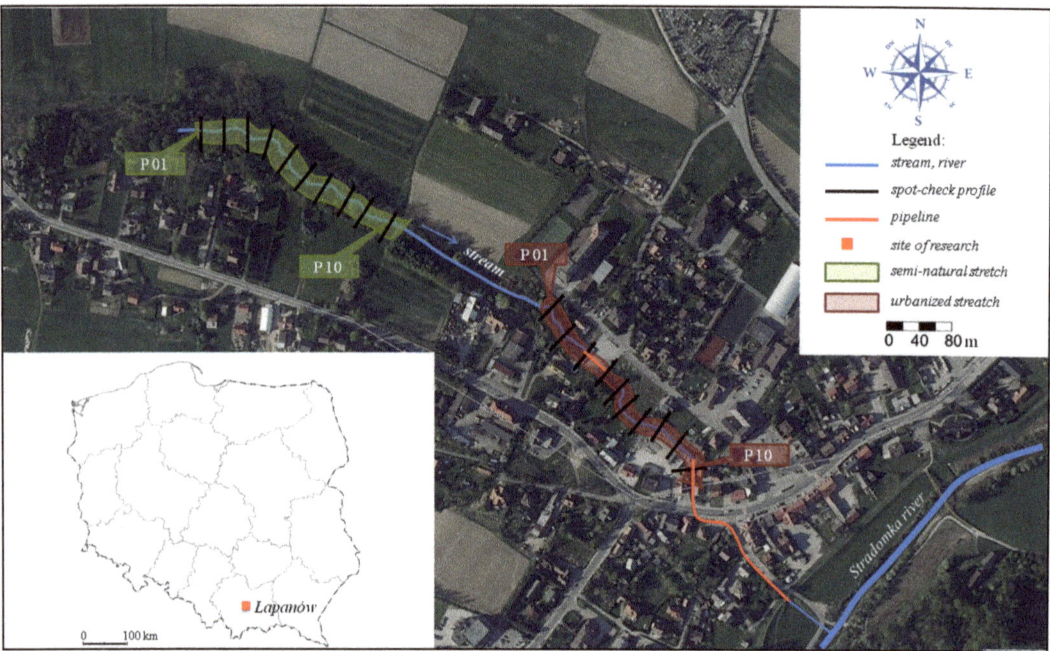

Figure 1. Location of the studies and sections with control profiles on the stream.

Based on the value of the RPI, the months and years were classified as follows: $RPI < 25$—extremely dry; $25 \leq RPI < 50$—very dry; $50 \leq RPI < 75$—dry; $75 \leq RPI \leq 125$—normal; $125 < RPI \leq 150$—wet; $150 < RPI \leq 200$—very wet; $RPI > 300.0$—extremely wet [34].

The thermal classification of months, seasons or years was prepared according to Lorenc [35,36] in relation to the average monthly temperature (T) and the average monthly temperature calculated over the multi-year period (T_{av}), increased or reduced by the standard deviation value (δ) (Table 1).

Table 1. Thermal classification of months, seasons and years.

Month, Period, Year	Criteria
extremely cold	$T > T_{av} + 2.5\delta$
anomalously cold	$T_{av} + 2.0\delta < T \leq T_{av} + 2.5\delta$
very cold	$T_{av} + 1.5\delta < T \leq T_{av} + 2\delta$
cold	$T_{av} + \delta < T \leq T_{av} + 1.5\delta$
slightly cold	$T_{av} + 0.5\delta < T \leq T_{av} + \delta$
normal	$T_{av} - 0.5\delta < T \leq T_{av} + 0.5\delta$
slightly warm	$T_{av} - \delta < T \leq T_{av} - 0.5\delta$
warm	$T_{av} - 0.5\delta < T \leq T_{av} - \delta$
very warm	$T_{av} - \delta < T \leq T_{av} - 1.5\delta$
anomalously warm	$T_{av} - 1.5\delta < T \leq T_{av} - 2\delta$
extremely warm	$T \leq T_{av} - 2.5\delta$

Explanations: T—air temperature in a given period; T_{av}—long-term average temperature in a given multi-year period; δ—standard deviation of air temperature in a given period. Source: elaborated according to Lorenc (2000).

A direct measurement of the water flow rate (Q) was performed using a bucket of known volume (V) and a stopwatch (t). The Q was calculated using the following formula:

$$Q = \frac{V}{t} \tag{2}$$

where: Q—water flow rate (m^3·s^{-1} or dm^3·s^{-1}); V—volume flow rate of water (m^3 or dm^3); and t—time (s).

The average annual water flow (SSQ) for small ungauged catchments located in the Carpathian region was calculated by Punzet formula [37–39]:

$$SSQ = 10^{-3} \cdot SSq \cdot A \tag{3}$$

$$SSq = 0.00001151 \cdot P^{2.05576} \cdot I^{0.0647} \cdot N^{-0.04435} \tag{4}$$

where: SSQ—average annual water flow (m^3·s^{-1}); SSq—average annual surface runoff (dm^3·s^{-1}·km^{-2}); A—catchment area (km^2); P—average annual precipitation (mm); I—river slope indicator (‰); and N—soil imperviousness index (%).

The soils in the catchment area were comprised of regoliths and loess-like dusty formations formed as a result of flysch weathering and simultaneous eolian sedimentation.

The physiographic characteristics of the catchment area included the 15 parameters [40,41] listed in Table 2, which presents: catchment geometry and shape, catchment morphometry, and hydrological conditions. On a 1:10,000 vector topographic map were areas with the following forms of land use: arable land, grassland and wasteland, forests (tree-covered areas) and built-up areas. The share of individual land use forms was calculated based on the total areas.

Table 2. Characteristics describing the stream catchment physiographic conditions.

Item	Parameter
1	Topographic catchment area A [km^2]—determined based on the planimetry of the area, which is closed by the boundary determining the topographic watershed
2	Maximum catchment length $L_{max.}$ [km]—main river valley length from the mouth to the point on the watershed in the extension of the spring
3	Catchment perimeter P [km]—length of the catchment's topographic watershed
4	Catchment mean width B_z [km]—ratio of the catchment topographic area (A) to the catchment length ($L_{max.}$): $B = \frac{A}{L_{max.}}$ [km]
5	Form coefficient C_f [–]*—ratio of the catchment area (A); ratio of the catchment area squared ($L_{max.}^2$): $C_f = \frac{A}{(L_{max.})^2}$
6	Elongation coefficient C_W [–]—ratio of the diameter of the circle of surface area equal to the catchment area, to the catchment length (L): $C_w = \frac{2}{L_{max.}} \cdot \sqrt{\frac{A}{\pi}}$
7	Circularity coefficient C_K [–]—ratio of the catchment area (A), to the surface area of the circle of the same perimeter as the catchment perimeter (P): $C_k = \frac{4 \cdot \pi \cdot A}{P^2}$
8	Gravelius catchment compactness coefficient GC [–]—ratio of the catchment perimeter (O_z), to the perimeter of the circle of the same surface area as the catchment area (A): $GC = \frac{P}{2 \cdot \sqrt{\pi \cdot A}}$

Table 2. *Cont.*

Item	Parameter
9	Minimum H_{min} and H_{max} [m a.s.l.]—taken from the topographic map as the lowest and the highest value on the watershed
10	Altitude H [m a.s.l.]—taken from the topographic map as the lowest altitude ($H_{min.}$) and the highest altitude ($H_{max.}$) of the spring ($H_{s.}$) and mouth ($H_{m,}$) on the watershed
11	Channel mean altitude H_{mean} [m a.s.l.]—arithmetic mean of the maximum altitude (H_{max}) and the minimum altitude (H_{min}) in the catchment area: $H_{mean} = \frac{H_{max} + H_{min}}{2}$ and from the Reitz formula: $H_{mean} = 0.434 \cdot \frac{H_{max} - H_{min}}{\log H_{max} - \log H_{min}}$
12	Catchment area denivelation (ΔH)—difference between the maximum (H_{max}) and minimum (H_{min}) altitude in the catchment area: $\Delta H = H_{max} - H_{min}$ [m]
13	Stream gradient/slope [%]—ratio of the altitude difference between the watercourse spring (H_s m a.s.l.) and mouth (H_m m a.s.l to the watercourse length in this section (L_s in m): $S = \frac{H_s - H_m}{L_s} \cdot 100$ [%]
14	Stream length L_s (km)—distance from the spring to the mouth
15	Land use types in catchment (%)—percentages of the various forms of use in the catchment area as arable land, forests etc.

Explanations: * [–]—dimensionless indicator/parameter.

The stream, at a length of about 250 m, carries water in the pipeline before spilling into the Stradomka river, a consequence of the progressive residential and service developments from the mid-1980s. In the middle part of the stream, a car park was built, which was also associated with the transformation of the water flow from an open stream bed into a closed pipeline. The increasing encroachment of urban areas in the stream catchment will follow the adopted land-use plan.

The hydromorphological evaluation of the 'unnamed stream' was performed in August 2020 with the use of the Hydromorphological Index for Rivers (HIR), which comprised of a field evaluation on two representative 300 m sections—semi-natural and urbanised—in which 10 control profiles (marked as P0–P10) were delimited about 30 m apart from each other, as well as a synthetic evaluation of the entire stream section (Figure 1). The study involved the determination of the presence and diversity of the natural elements of the stream and of the valley, and characterisation of the range of modifications in the watercourse's morphology [30,42].

The land use in the stream valley was determined based on the orthophotomap from 2020, available at geoportal.gov.pl (accessed on 20 December 2021) (supported by a site visit), within a 100 m wide buffer on each bank; as a percentage of the area and then used to determine the weight coefficient (w) to calculate the mean HIR. Out of three basic land use forms, i.e., urbanised (U), agricultural (R) and semi-natural (S), the urbanised areas dominated in the buffer—sealed surfaces with dense or dispersed development, roads and other anthropogenic non-built-up areas, which made up about 58% of the buffer area (wU = 69.9). Next, there were semi-natural areas—forests, trees and shrubs, marshy areas, rush and herb vegetation, covering about 25% of the buffer area (wS = 30.1). The remaining 17% were agricultural areas—arable land, grassland and allotment gardens (wA = 0.0). Only the land use forms with a share $\geq 25\%$ in the buffer were used in the calculation [42]. Two study sections were determined based on the land use analysis in the stream valley: the first in the semi-natural part, and the second in the urbanised part. Due to the stream features (its small length and partial pipelining) and the property conditions in the catchment area (the stream flows through private properties, often fenced), the study sections lengths were

300 m long instead of the standard 500 m, which is acceptable according to the authors of the method [30,42].

For each stretch, the HIR value was calculated. The multimetric HIR index combined two indices: The Hydromorphological Diversity Score (HDS) and the Habitat Modification Score (HMS). The HDS informed about the presence of natural attributes in the channel, coastal zone and the river valley. Each of the HDS attributes delivered a range of points, enabling the calculation of the HDS of the river stretch. HMS provided information on the hydromorphological modifications; it included various forms of fluvial ecosystem transformations, such as profile modifications and reinforcements, and the presence and abundance of engineering facilities. The detailed procedure of scoring is presented in the HIR method manual. In practice, HDS and HMS values usually do not exceed 100 [30,42].

$$HIR = \frac{\left(\frac{HDS - HMS}{100}\right) + 0.85}{1.8} \tag{5}$$

The obtained values were compared with the classes for a given type of watercourse: class I $HIR \geq 0.824$; class II $HIR \geq 0.715$; class III $HIR \geq 0.600$; class IV $HIR \geq 0.485$; and class V $HIR < 0.485$.

The assessment of the hydromorphological state of the stream catchment consisted of calculating the weighted average of the HIR values, taking into account the weighting factor for three forms of land development (calculated on the basis of their percentage share in the buffer), according to the following formula:

$$HIR_{mean} = \frac{(HIR_U \cdot wU) + (HIR_A \cdot wA) + (HIR_S \cdot wS)}{100} \tag{6}$$

where: HIR_{mean}—average HIR value for the whole water body; HIR_U—HIR calculated for the survey sites located in an urbanised area; wU—weighting factor for urbanised areas calculated on the basis of their percentage share in the buffer; HIR_A—HIR calculated for the survey sites located in an agricultural area; wA—weighting factor for agricultural areas calculated on the basis of their percentage share in the buffer; HIR_S—HIR calculated for the survey sites located in a semi-natural area; and wS—weighting factor for semi-natural areas calculated on the basis of their percentage share in the buffer.

3. Results

3.1. Meteorological and Hydrological Characteristics

The characteristics of the meteorological conditions are presented in Figures 2 and 3. The climate is fairly dry in the winter half-year, and is characterised by fairly hot summers, where short rain showers are quite common and often come in intense bursts.

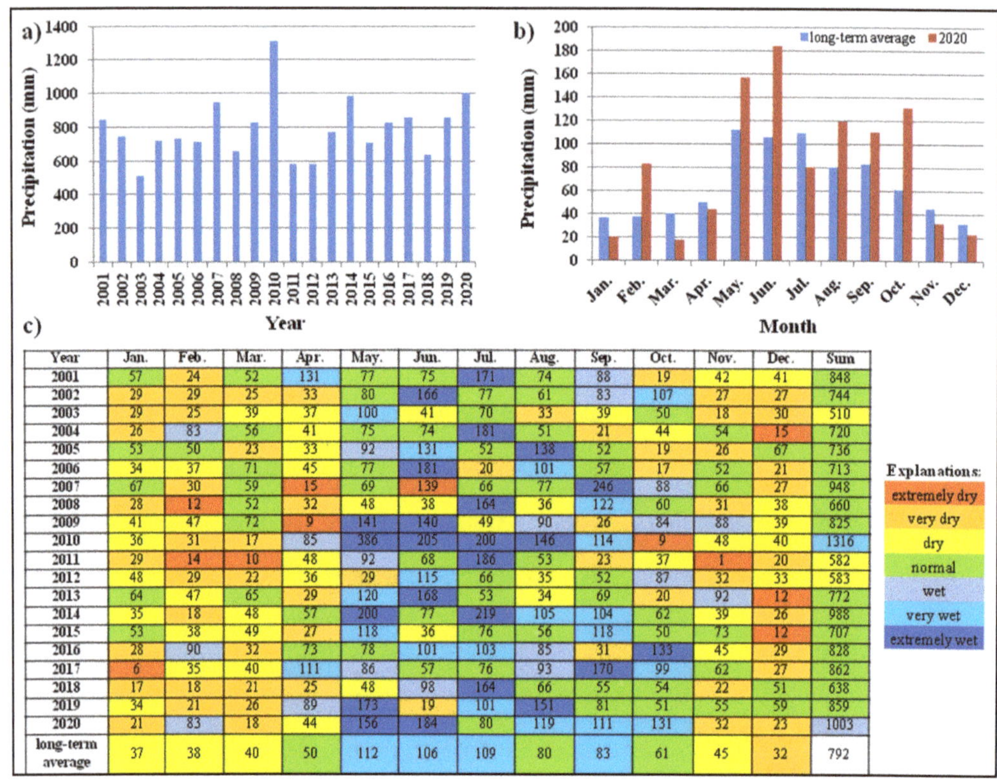

Year	Jan.	Feb.	Mar.	Apr.	May.	Jun.	Jul.	Aug.	Sep.	Oct.	Nov.	Dec.	Sum
2001	57	24	52	131	77	75	171	74	88	19	42	41	848
2002	29	29	25	33	80	166	77	61	83	107	27	27	744
2003	29	25	39	37	100	41	70	33	39	50	18	30	510
2004	26	83	56	41	75	74	181	51	21	44	54	15	720
2005	53	50	23	33	92	131	52	138	52	19	26	67	736
2006	34	37	71	45	77	181	20	101	57	17	52	21	713
2007	67	30	59	15	69	139	66	77	246	88	66	27	948
2008	28	12	52	32	48	38	164	36	122	60	31	38	660
2009	41	47	72	9	141	140	49	90	26	84	88	39	825
2010	36	31	17	85	386	205	200	146	114	9	48	40	1316
2011	29	14	10	48	92	68	186	53	23	37	1	20	582
2012	48	29	22	36	29	115	66	35	52	87	32	33	583
2013	64	47	65	29	120	168	53	34	69	20	92	12	772
2014	35	18	48	57	200	77	219	105	104	62	39	26	988
2015	53	38	49	27	118	36	76	56	118	50	73	12	707
2016	28	90	32	73	78	101	103	85	31	133	45	29	828
2017	6	35	40	111	86	57	76	93	170	99	62	27	862
2018	17	18	21	25	48	98	164	66	55	54	22	51	638
2019	34	21	26	89	173	19	101	151	81	51	55	59	859
2020	21	83	18	44	156	184	80	119	111	131	32	23	1003
long-term average	37	38	40	50	112	106	109	80	83	61	45	32	792

Figure 2. The characteristics of the precipitation conditions: (**a**) average annual precipitation in 2001–2020; (**b**) average monthly precipitation from the multi-year period 2001–2020 and in 2020; (**c**) classification of precipitation conditions on the basis of Relative Precipitation Index (RPI) for the months and years.

Typically, 30% of the precipitation falls during the winter half-year and 70% during the summer half-year. In the multi annual period from 2001–2020, the average annual air temperature was 9.1 °C, and total precipitation was 792 mm. Increased average temperatures and rainfall are observed over the analysed period, which is in line with a report by the Institute of Meteorology and Water Management—National Research Institute, which has been monitoring Poland's climate for over 100 years on an ongoing basis [43]. To assess the meteorological conditions of the region where the study was conducted, the relative precipitation index (RPI) values were calculated, as a measure of the precipitation efficiency in a given month and year. Based on the RPI values calculated for the months from 2001–2020, as shown in Figure 2c, it was determined that 112 of the months were dry, 63 of the months were optimal and 65 of the months were wet. In terms of precipitation, 2020 belonged to the wet year. The average multi annual temperature varies between 7.8 °C and 10.3 °C (Figure 3c). The water temperature depends on the air temperature which, in turn, translates into the amount of oxygen in the water [44]. The annual average temperature in 2020 was 0.7 °C higher than the multi annual average. In the analysed multi-year period, two years were very cold, six years were slightly cold, six years were normal, two were slightly warm, three years were warm and one year was very warm. In general, the variable meteorological conditions in the study area, characterised by periods of excesses and shortages of water, indicate the need for good maintenance of the hydromorphological conditions of watercourses, mainly drainage functions.

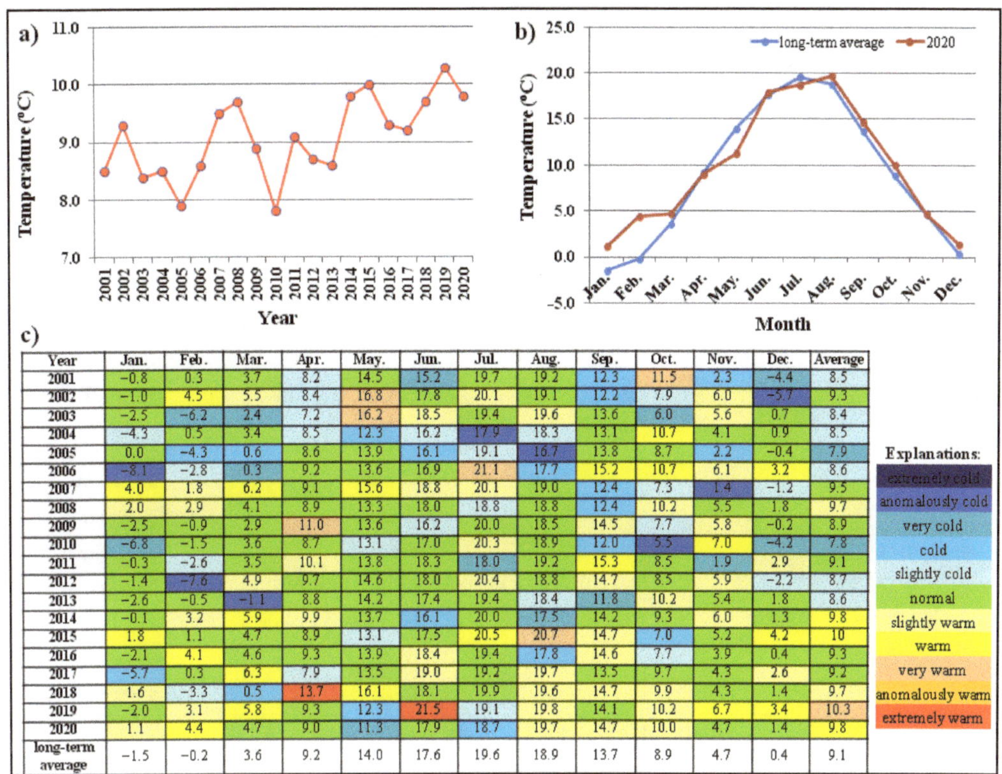

Figure 3. The characteristics of the thermal conditions: (**a**) average annual temperature in 2001–2020; (**b**) average monthly temperature from the multi-year period 2001–2020 and in 2020; (**c**) monthly thermal classification proposed by Lorenc (2000).

Year	Jan.	Feb.	Mar.	Apr.	May.	Jun.	Jul.	Aug.	Sep.	Oct.	Nov.	Dec.	Average
2001	−0.8	0.3	3.7	8.2	14.5	15.2	19.7	19.2	12.3	11.5	2.3	−4.4	8.5
2002	−1.0	4.5	5.5	8.4	16.8	17.8	20.1	19.1	12.2	7.9	6.0	−5.7	9.3
2003	−2.5	−6.2	2.4	7.2	16.2	18.5	19.4	19.6	13.6	6.0	5.6	0.7	8.4
2004	−4.3	0.5	3.4	8.5	12.3	16.2	17.9	18.3	13.1	10.7	4.1	0.9	8.5
2005	0.0	−4.3	0.6	8.6	13.9	16.1	19.1	16.7	13.8	8.7	2.2	−0.4	7.9
2006	−8.1	−2.8	0.3	9.2	13.6	16.9	21.1	17.7	15.2	10.7	6.1	3.2	8.6
2007	4.0	1.8	6.2	9.1	15.6	18.8	20.1	19.0	12.4	7.3	1.4	−1.2	9.5
2008	2.0	2.9	4.1	8.9	13.3	18.0	18.8	18.8	12.4	10.2	5.5	1.8	9.7
2009	−2.5	−0.9	2.9	11.0	13.6	16.2	20.0	18.5	14.5	7.7	5.8	−0.2	8.9
2010	−6.8	−1.5	3.6	8.7	13.1	17.0	20.3	18.9	12.0	5.5	7.0	−4.2	7.8
2011	−0.3	−2.6	3.5	10.1	13.8	18.3	18.0	19.2	15.3	8.5	1.9	2.9	9.1
2012	−1.4	−7.6	4.9	9.7	14.6	18.0	20.4	18.8	14.7	8.5	5.9	−2.2	8.7
2013	−2.6	−0.5	−1.1	8.8	14.2	17.4	19.4	18.4	11.8	10.2	5.4	1.8	8.6
2014	−0.1	3.2	5.9	9.9	13.7	16.1	20.0	17.5	14.2	9.3	6.0	1.3	9.8
2015	1.8	1.1	4.7	8.9	13.1	17.5	20.5	20.7	14.7	7.0	5.2	4.2	10
2016	−2.1	4.1	4.6	9.3	13.9	18.4	19.4	17.8	14.6	7.7	3.9	0.4	9.3
2017	−5.7	0.3	6.3	7.9	13.5	19.0	19.2	19.7	13.5	9.7	4.3	2.6	9.2
2018	1.6	−3.3	0.5	13.7	16.1	18.1	19.9	19.6	14.7	9.9	4.3	1.4	9.7
2019	−2.0	3.1	5.8	9.3	12.3	21.5	19.1	19.8	14.1	10.2	6.7	3.4	10.3
2020	1.1	4.4	4.7	9.0	11.3	17.9	18.7	19.7	14.7	10.0	4.7	1.4	9.8
long-term average	−1.5	−0.2	3.6	9.2	14.0	17.6	19.6	18.9	13.7	8.9	4.7	0.4	9.1

The average annual water flow of the unnamed stream is placed at $SSQ = 0.004 \ \mathrm{m^3 \cdot s^{-1}}$ and corresponds to the flow measured directly ($Q = 0.005 \ \mathrm{m^3 \cdot s^{-1}}$). The catchment basin of the stream receives water both from rainfall and from snow melt in the slope. The higher precipitation levels in 2020 lead to a higher supply of water to the stream. Based on conversations with residents, there have been dry years or months in which the stream did not carry water.

3.2. Catchment Characteristics

The basic physiographic parameters describing the catchment of the examined stream are presented in Table 3. The stream catchment area is very small (<10 km²). The catchment length is just over 1 km, its mean width is 0.23 km, and the perimeter is about 3 km. The form coefficient (*Cf*), elongation coefficient (*Cw*), circularity coefficient (*Ck*) and compactness coefficient (*Cz*) indicate that the stream catchment is narrow and elongated. The denivelation of the catchment area is 59 m, with a watershed slope of about 23.9‰. The mean altitude in the catchment is about 258 m a.s.l.

Table 3. Physiographic parameters of the studies stream catchment.

Parameter		Value
Catchment Geometry		
Topographic catchment area (A):		0.31 km^2
Maximum catchment length ($L_{max.}$):		1.37 km
Catchment perimeter (P):		3.10 km
Mean catchment width (B_z):		0.23 km
Shape coefficient:	form (Cf):	0.17
	elongation (Cw):	0.46
	circularity (Ck):	0.41
	Gravelius compactness (GC):	1.57
Catchment Morphometry		
Altitude:	minimum/mouth ($H_{min.}$):	229.10 m a.s.l.
	maximum ($H_{max.}$):	288.30 m a.s.l.
	spring ($H_{źr.}$):	253.00 m a.s.l.
	mean ($H_{śr.}$):	258.70 m a.s.l.
	mean ($H_{śr.}$) acc. to Reitz:	257.44 m a.s.l.
Denivelation (ΔH):		59.20 m
Stream gradient/slope (Jc):		2.4% (23.9‰)
Catchment Hydrography		
Stream length (Lc):		~1.00 km
Type of Land Use		
Forests		6%
Arable land		41%
Grassland		28%
Built-up areas		25%
Sum		100%

The dominant land use is arable land with slightly over 40% of the catchment area. The catchment afforestation rate is 6%. The trees and shrubs occur mainly in the stream valley, thus forming the natural biological development of the watershed bed. The urbanised area (dispersed and dense development) is 25%. Grassland (meadows/pastures) and wasteland with grass vegetation form about 28% of the catchment area (Table 3, Figure 4).

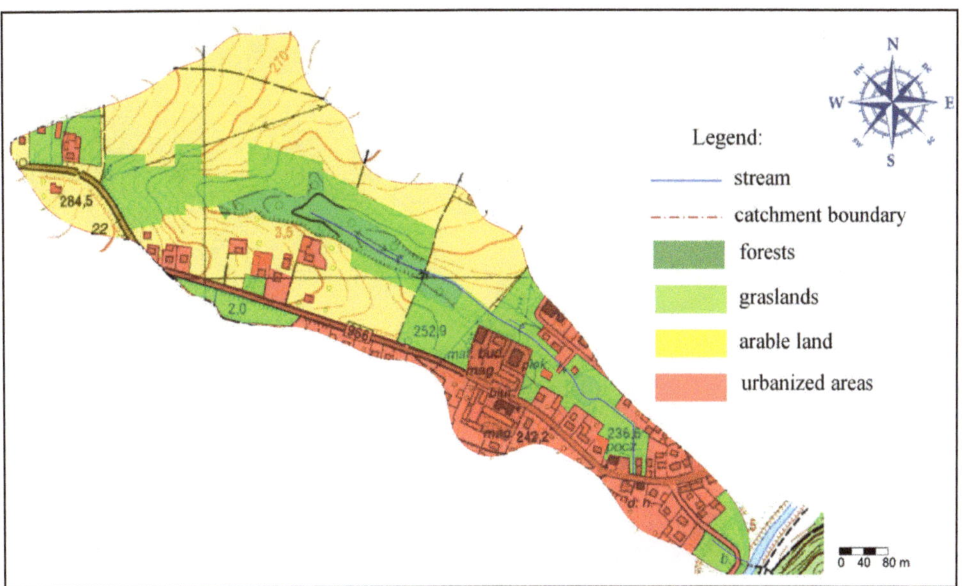

Figure 4. Land use structure in the stream catchment.

3.2.1. Characteristics of the Semi-Natural Section

Marshy areas without a distinctively formed bed can be found near the control point P01—the spring area. Valuable environmental elements include marshy meadows, as a spring area in the form of a weak outflow of groundwaters to the surface (seeps, bogs). The dominant land use forms include forest, trees/shrubs and grassland (marshy meadows) (Figure 5A). From cross-section P02, the bed-bottom material is classified as clay and silt, sometimes covered with a thin layer of mud (Figure 5B). The average water depth is about 3.0 cm. No modifications in the bed and natural morphologic elements were found. Soil (loess) is deposited on the slopes. The banks' cross-sections in the majority of the control profiles are eroded (as a result of water erosion), with vertical or underwashed banks of the stream. No modifications were found on the slopes, and the natural morphological elements include eroding bank undercuts with visible plant roots and numerous bank outwashes not fixed with vegetation (Figure 5C). There is no vegetation in the bed, and the vegetation appears on the slope as a uniform structure (Figure 5C) with ferns and grass vegetation, and as a complex structure on the slope tops with continuous tree stands (mainly maple, alder, oak, and spruce and grass vegetation). The stream valley in the semi-natural section is very shaded (50–75%). The land use in the areas near the banks is forest, trees and shrubs. The synthetic evaluation of the entire examined section (300 m) indicates an unobstructed flow in the bed, despite an occasional weak natural swelling by minute wood bed load (Figure 5D) and the vegetation in the bed and on the slope being typical of marshy areas (i.a. sedge). Detritus in the form of small-size dead organic matter from falling leaves and weak lateral tributaries can be found in the upper course of the stream. The laminar water flow dominates in the bed (this was the case when the studies were conducted, but rapid flow and overflow was also observed, mainly upstream in places where erosion faults/incisions formed small waterfalls). Florae were found in the stream valley, in the form of ivy (*Hedera helix*) and small balsam (*Impatiens parviflora* DC.) Faunae were also found, in the form of Roman snail (*Helix pomatia*) and slug. Debris classified as refuse/waste was also found the stream valley.

Figure 5. Selected elements of the hydromorphological diversity index in the semi-natural section: (**A**)—marshy areas and spring zone; (**B**)—laminar water flow; uniform vegetation on the slopes; (**C**)—eroding bank undercuts; bank outwash not fixed with vegetation; (**D**)—natural swelling by small wood bed load.

3.2.2. Hydromorphological Characteristics of the Urbanised Section

In the initial control profiles of the urbanised section (P01–P03), the dominant bed material is fine sand and dust covered by a layer of muddy sediments. The water flow in the bed is laminar (Figure 6A). The water level in the bed is about 6 cm. The left-hand and right-hand banks, in the 1 m wide strip, are made of loose material (soil). In the 5 m strip from the bank, there are semi-continuous trees and shrubs, and a simple vegetation structure on the top and slopes of the banks; these are mainly alder (*Alnus*), maple (*Acer*), oak (*Quercus*) and spruce (*Picea abies*); there is also moss (*Bryophyta*), particularly on the tree branches and ferns (*Polypodiopsida*). No water plants were found in the bed. The bank profile ranges from gentle, which is classed as <45° (left-hand bank), to steep, which is >45° (right-hand bank). Typical land use in the strip 50 metres away from the bank top was trees/shrubs and buildings (school, private houses). Shading of the bed is visible, as well as exposed roots and outgrowth on the slopes of both banks. The natural erosion-related morphological elements include an undercut on the right-hand bank.

Figure 6. Selected elements of the hydromorphological diversity index in the urbanised section: (**A**)—overgrown bed with laminar flow; (**B**)—outlet of Ø 1.00 m concrete culvert; (**C**)—grass vegetation (wasteland) and built-up areas; (**D**)—trees and shrubs. Grass vegetation (wasteland).

Farther in the urbanised section, the dominant bed-bottom materials are still fine sand and silt. The water flow is laminar. For about 30 metres, the stream flows in the pipe (the closed bed) with a paved surface over it (the car park). Then, the water flows out of the Ø = 1.00 m concrete pipe (the culvert) (Figure 6B), and the flow is $Q \approx 0.005 \text{ m}^3 \cdot \text{s}^{-1}$. There are no cross-section modifications in the open bed, and no natural morphological elements were found in the bed. The material of the right-hand and left-hand slopes is soil (sand and dust are dominating). No modifications on the right-hand and left-hand slopes, and no natural morphological elements were found. The bed is heavily overgrown with vegetation; these are mainly plants above the water surface or recumbent amphiphites. In section P04–P07, the watercourse runs through uncultivated land. There is simple vegetation structure on the right-hand bank in the form of single trees/shrubs—mainly willow (*Salix*), hazel (*Corylus*) and oak (*Quercus*)—and green vegetation—mainly reed (*Phragmites australis*) and sedge (*Carex*) (Figure 6C,D). Uniform vegetation—mainly grass vegetation (*Phragmites australis*)—and tall herb vegetation—mainly nettle (*Urtica dioica*)—dominate on the left-hand slope. Vegetation typical of marshy habitats and areas has also been found, i.e., purple loosestrife (*Lythrum salicaria*) and marsh horsetail (*Equisetum palustre*). The plot through which the stream flows is built up/developed on both sides. There is a road to the north with housing development along it, and single-family houses to the east (Figure 6C). Refuse/waste was found near the bank on the south side, behind the fence.

From profile P07 to P10, the stream bed is heavily overgrown with grass vegetation, with clear signs of anthropopression in the form of bottom- and slope- grading. The dominant bed-bottom material is fine sand. There is soil on the slopes. The water flow in the bed is laminar. Farther along, the stream is pipelined (concreted) and flows under the urban development; before flowing into the Stradomka river, the stream bed is open again with visible signs of grading and vegetation mowing (inter-embankment zone of the Stradomka river). In addition, from among the elements recorded in the synthetic evaluation, gullies and culverts were found. The bed dimension for the semi-natural and urbanised section are presented in Figure 7A,B, respectively.

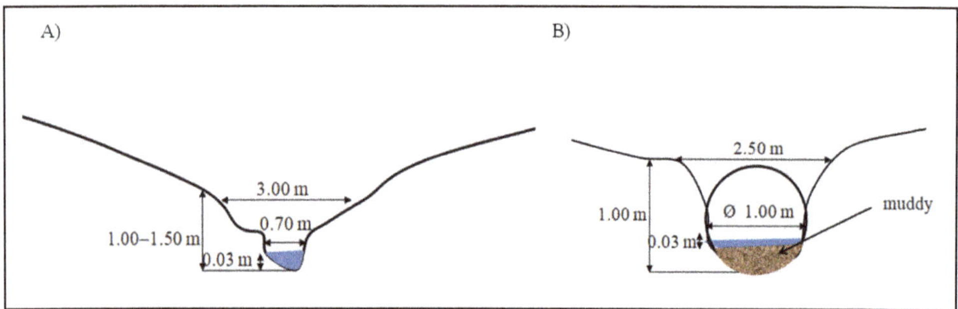

Figure 7. Bed dimensions: (**A**)—semi-natural section; (**B**)—urbanised section.

The HDS index indicates the hydromorphological diversity of the stream. The HDS values calculated for the individual study sections are varied from 27 for the urbanised section to 56 for the semi-natural section (Table 4). The HDS values for the semi-natural section are mainly influenced by: the presence of the natural morphological elements of the slopes in the control profiles (17.8%); the diversity of elements accompanying the trees (14.3%); and the naturalness and heterogeneity of the stream valley use (14.3%). For the urbanised section, the highest percentage of HDS value is found in the bed material heterogeneity (15.0%), the natural morphological elements of the slopes (15.0%), the vegetation diversity in the channel bed (15.0%), and the structure of the bank vegetation (15.0%). The HMS values reflecting the degree of anthropogenic changes in the hydromorphology of streams and rivers indicate that the lower part of the examined stream is the most modified (Table 4). The score for the lower-urbanised stream section is 15.5, indicating a significant modification of the stream related mostly to the modifications in control profiles (38.7%), especially profiling the bottom and slopes of the stream and the disturbance of the connectivity with the river valley (25.8%); the latter is especially influenced by urban areas. The upper examined section, for which the HMS is 0, is a hydromorphological habitat which has not been significantly modified. It is located within the tree stand boundaries and remains almost in semi-natural conditions.

It is unequivocally found that the value of the HIR index varies significantly for the tested sections. The final HIR value for the semi-natural section is 0.78, and is 44% higher in relation to the urbanised section (0.54); the changes in its numerical value mainly depend on the heterogeneity of the water flow, the natural morphological elements of the slopes, the diversity of elements accompanying the trees, and the natural land use of the valley (Tables 4 and 5). According to the classification [30], the hydromorphological condition in the semi-natural section is determined to be good, and in the urbanized section, to be poor. The hydromorphological condition of the whole stream is determined to be moderate (Table 5).

Table 4. Components of the Hydromorphological Diversity Score (HDS) and the Hydromorphology Modification Score (HMS).

Parameters Collected during the On-Site Evaluation	Section	
	Semi-Natural	Urbanised
HDS		
1. The stream channel zone—riverbed		
1.1. Variation of the river line	3.0 (5.3%)	0.0 (0.0%)
1.2. Variation of the riverbed slope	3.0 (5.3%)	1.0 (3.5%)
1.3. Heterogeneity of water flow	5.0 (9.0%)	3.0 (11.0%)
1.4. Bed material heterogeneity	4.0 (7.3%)	4.0 (15.0%)
1.5. Natural morphological elements of the bed-bottom	1.0 (1.8%)	0.0 (0.0%)
1.6. Natural morphological elements of the slopes	10.0 (17.8%)	4.0 (15.0%)
1.7. Vegetation diversity in the bed	5.0 (9.0%)	4.0 (15.0%)
2. The stream channel zone—bank face		
2.1. Vegetation structure on the slopes	0.0 (0.0%)	0.0 (0.0%)
2.2. Diversity of elements accompanying the trees	8.0 (14.3%)	1.0 (3.5%)
3. The stream valley adjacent to the bank-top zone		
3.1. Structure of bank-top vegetation	5.0 (9.0%)	4.0 (15.0%)
3.2. Not-managed bank-top zone	4.0 (7.3%)	3.0 (11.0%)
4. The stream valley zone		
4.1. Natural land use of the valley	8.0 (14.3%)	3.0 (11.0%)
4.2. Connection between the stream and the valley	0.0 (0.0%)	0.0 (0.0%)
ΣHDS	56.0 (100%)	27.0 (100%)
HMS		
1. Transformed transverse section of the stream channel	0.0 (0.0%)	2.0 (12.9%)
2. Hydroengineering structures	0.0 (0.0%)	0.5 (3.2%)
3. Transformations observed in spot-checks	0.0 (0.0%)	6.0 (38.7%)
4. Disturbance of the connectivity with the river valley	0.0 (0.0%)	4.0 (25.8%)
5. Other types of human degradation	0.0 (0.0%)	3.0 (19.4%)
ΣHMS	0.0 (0.0%)	15.5 (100%)

Table 5. Results of hydromorphological evaluation and quality classes of individual examined sections and the whole stream.

Item	Measuring Section	Study Section			Watercourse/Stream (Whole)		
		HIR	Hydromorphological Condition	Class	HIR_{mean}	Hydromorphological Condition	Class
1	semi-natural	0.78	good	II	0.61	moderate	III
2	urbanised	0.54	poor	IV			

4. Discussion

In the catchment area of the analyzed stream, an increase in the sum of annual precipitation and the average annual temperature is visible. Carpathian streams play an important role in draining the local catchment area, especially during flash floods, which are appearing more and more frequently [45]. The highest flood risks in Poland are characteristics of the Carpathian tributaries of the upper Vistula river, including the catchment area of the studied stream [46]. The negative impact of extreme weather phenomena is already visible, with drought and soil erosion bringing major problems, especially in the Carpathian Region; however, as reported by Ionita et al. [47], the frequency of droughts has not unusually increased in Central Europe when compared to preindustrial drought records. Natural climate change has led to changes in the water regimes of small upland and mountain streams, thus causing the occurrence of flash floods, or the drying up of watercourses in late summer [48]. The results of climate models show that in the area of Poland, in the near and far future, there will be an increase in air temperatures and in precipitation [49–51]. On the other hand, according to Atwood et al. [52], Climate models are quite poor when it comes to rainfall on a regional level. Any future trends in rainfall coming from climate models have to be treated with utmost caution. Attempts to identify climate change in the city micro-environment and improvement through water retention are conducted in Slovakia [53].

The physiographic (geometrical and morphometric) parameters of the catchment indicate a mountainous character of the examined stream; this includes a small area and significant elongation, a mean elevation of >200 m a.s.l., and a large slope [54]. The results of hydromorphological evaluation revealed significant differences between the semi-natural and urbanised part of the stream, indicating regulation and canalisation as the main reasons for hydromorphological degradation. Studying the catchment areas of two Carpathian flysch streams (the Jaszcze and Jamne), Bucała-Hrabia and Wiejaczka [54] concluded that a higher diversity of hydromorphological forms, with a corresponding HDS, occurs in the upper and medium course of the streams in the upland landscape; they also concluded that a lower diversity occurs in the lower part of the streams' catchment areas, which are subjected to anthropopression. A similar regularity of naturality and anthropogenic changes of the habitat between the upper and lower course of streams in Poland, India and China was shown by Kijowska-Strugała et al. [55].

The results of the scoring correspond to the preliminary hypothesis, according to which, despite the small stream, there is a noticeable difference between the analysed sections of the stream. An extensive amount of literature on linkages between the effects of anthropopression in stream habitats and river hydromorphology reports a number of relationships, at many levels. The need is urgent to develop refined and updated hydromorphological assessment systems targeting small ungauged watercourse evaluation, for use by the European Water Framework Directive (WFD) and national water-related policies. The topography and rich vegetation in the river's catchment area may make it difficult to perform field studies. Recent studies shows that Unmanned Aerial Vehicles (UAVs) can be used for inventory and assessment of the shydromorphological status of streams and rivers, especially in hard-to-reach areas. Drones and digital photogrammetry now provide an alternative approach for monitoring river habitats and hydromorphology [56,57].

The aim of the research currently carried out in Ireland is to advance knowledge on the role of small streams in water quality, biodiversity and ecosystem services protection; this will inform policy, measures and management options to meet water quality and other resources protection targets [58].

As a rule, natural watercourses have a higher HDS and the artificial watercourses (ditches, channels) have a higher HMS [21]. Hajdukiewicz et al. [6] indicate sbed regulation and significant grading (canalisation of the river) as the main causes of the hydromorphological degradation of the Biała river. Bryndal et al. [59] prove that high human activity in the catchment area deteriorates its natural drainage/water removal, thus contributing to a more frequent occurrence of high water stage and floods. Bedla et al. [60] and Pietruczuk, et al. [61] indicate that that the excessive piping of a river or stream adversely affects the natural environment in their valleys, deteriorating the aesthetic values of landscape. The diversity of morphological forms (flow type, bottom and slope material) plays a significant role in the biodiversity of fauna and flora in the watercourse bed and valley [28,62,63]. This can be seen in the examined stream, where the vegetation cover of the watercourse body is much larger in the lower urbanised part of the stream than in the upper, semi-natural part (Table 4), characterised by a great slope and a larger dynamic of the flowing water. In the studied stream, it was demonstrated that, depending on the vegetation diversity, the HDS variation could change by about 25%. Furthermore, Kiraga [64] reported that, depending on the season, the vegetation diversity variation could change, which could lead to a step from one hydromorphological class to another. It was similar in the case of the HMS parameter, where modifications in the riverbed and stream valley were visible in the urbanised section. These changes significantly reduced the hydromorphological condition. It is well known that riparian vegetation plays a crucial role in sustaining river hydromorphological conditions [65]. Forests, trees and shrubs play an important part in the hydromorphological diversity of the zone near the stream banks. They act as biological protection (natural development) of the stream bed and valley, increasing the catchment retention, equalising the flows, and slowing down erosion of the stream slopes, bottom and banks. The root systems of these plants protect the soil from being washed out. Kałuża et al. [66] reported that, in the forested river reach, bottom and riverbank vegetation was completely absent. The river channel was narrow; however, due to the vicinity of trees and shrubs, considerable accumulation of organic matter was observed in the river channel. The flora acted as basic protection, mainly in the upper part of the catchment, where the land gradient was higher [62,67]. The majority of soils in the Wiśnicz Foothills, including the soils in the catchment of the examined stream, are made of Carpathian flysch (loess) which, without appropriate vegetation cover, are subject to intensive erosion [1,54].

Any transformation of the stream channel (riverbed) and catchment area should be thought out and included in the land use plan, as some of them might prove to be irreversible and have a destructive impact on the quality of the water environment [68–70].

5. Conclusions

The conclusions of the studies are as follows:

In its upper course, the stream has winding meanders, where the banks are subjected to erosion by flowing water, particularly during high water and high flow speed. In addition, it has low anthropopression and moderate diversity of morphological forms and conditions. Its lower course, on the other hand, has been significantly modified and transformed, in both the bed and the banks. The mouth part of the stream belongs to the intensively modified part of the surface waters. Factors significantly worsening hydromorphological conditions are various forms of anthropopressure, with particular emphasis on urbanization.

The evaluation indicates hydromorphological features of the stream which have been changed significantly at canalised sections, and which will most likely undergo the largest improvement when the bank protections are removed and the free channel migration is made possible, which—as a result of the progressing development of the lower part of the stream—is no longer possible.

This applied method can be used for the hydromorphological assessment and inventory of small, ungauged streams, because its scope includes elements that testify to the naturalness of stream habitats and a radical transformation in stream catchments. A major advantage of the HIR method is that it accounts for the HDS and HMS; this allows one to determine to what extent a given watercourse section is natural, and to what extent it is transformed. On the other hand, the final valorisation results are strongly affected by the presence of vegetation, which varies during the whole year. The type of water flow may also turn out to be an insufficient parameter—especially in periodic streams or those susceptible to rapid fluctuations in the water level after heavy rainfall.

This example of an examined stream indicates the need for more frequent monitoring the catchment areas of small streams not included in any hydrological classifications.

The authors believe that proper management of water resources in small, ungauged catchments lies with the local and regional authorities. All actions taken in the stream bed, and its catchment area, should be in accordance with the principle of sustainable development.

Based on this hydromorphological assessment, the results obtained helped us to evaluate the environmental changes and anthropogenic pressures on the stream sections; however, further research is required on the changes in the hydromorphological status of small watercourses.

Author Contributions: Conceptualization, Ł.B. and T.K.; methodology, Ł.B. and T.K.; software, Ł.B.; validation, Ł.B.; formal analysis, Ł.B. and T.K.; investigation, Ł.B. and T.K.; resources, Ł.B. and T.K.; data curation, Ł.B. and T.K.; writing—original draft preparation, Ł.B.; writing—review and editing, Ł.B. and T.K.; visualisation, Ł.B.; supervision, T.K.; project administration, Ł.B.; funding acquisition, T.K. All authors have read and agreed to the published version of the manuscript.

Funding: This research was funded by the Department of Land Reclamation and Environmental Development, Faculty of Environmental Engineering and Land Surveying, University of Agriculture in Krakow.

Institutional Review Board Statement: Not applicable.

Informed Consent Statement: Not applicable.

Data Availability Statement: Not applicable.

Conflicts of Interest: The authors declare no conflict of interest.

References

1. Dawson, C.P.; Hendee, J.C. *Introduction to Forests and Renewable Resources*, 9th ed.; Waveland Press: Long Grove, IL, USA, 2019; Available online: https://books.google.pl/books?id=wZG4DwAAQBAJ&pg=PA430&lpg=PA430&dq=upland+stream+gradient&source=bl&ots=UMIyzqRIPy&sig=ACfU3U2VKor7jpnxwgScPDzliJYmtwLgOw&hl=pl&sa=X&ved=2ahUKEwjW74PP5L_0AhXilYsKHTexBLIQ6AF6BAg0EAM#v=onepage&q=upland%20stream%20gradient&f=false (accessed on 26 November 2021).
2. Głowacki, W. Newly Emerging Geosites in the Polish Western Outer Carpathians as an Asset for Geoeducation and Geotourism. *Geoheritage* **2019**, *11*, 1247–1256. [CrossRef]
3. Hauer, C.; Pulg, U. The non-fluvial nature of Western Norwegian rivers and the implications for channel patterns and sediment composition. *Catena* **2018**, *171*, 83–98. [CrossRef]
4. Hauer, C.; Pulg, U. Buried and forgotten—The non-fluvial characteristics of postglacial rivers. *River Res. Appl.* **2021**, *37*, 123–127. [CrossRef]
5. Kondracki, J. *Regional Geography of Poland*; PWN: Warszawa, Poland, 2011. (In Polish)
6. Hajdukiewicz, H.; Wyżga, B.; Zawiejska, J.; Amirowicz, A.; Oglecki, P.; Radecki-Pawlik, A. Assessment of river hydromorphological quality for restoration purposes: An example of the application of RHQ method to a Polish Carpathian river. *Acta Geophys.* **2017**, *65*, 423–440. [CrossRef]
7. Bryndal, T.; Kroczak, R. Reconstruction and characterization of the surface drainage system functioning during extreme rainfall: The analysis with use of the ALS-LIDAR data—the case study in two small flysch catchments (Outer Carpathian, Poland). *Environ. Earth Sci.* **2019**, *78*, 215. [CrossRef]
8. Jakubínský, J.; Pelíšek, I.; Cudlín, P. Linking Hydromorphological Degradation with Environmental Status of Riparian Ecosystems: A Case Study in the Stropnice River Basin, Czech Republic. *Forests* **2020**, *11*, 460. [CrossRef]
9. Bogdał, A.; Kowalik, T. Variability of values of physicochemical water quality indices along the length of the Iwoniczanka stream. *J. Ecol. Eng.* **2015**, *16*, 168–175. [CrossRef]
10. Policht-Latawiec, A.; Bogdal, A.; Kanownik, W.; Kowalik, T.; Ostrowski, K. Variability of physicochemical properties of water of the transboundary Poprad river. *J. Ecol. Eng.* **2015**, *16*, 100–109. [CrossRef]
11. Allaire, M.C.; Vogel, R.M.; Kroll, C.N. The hydromorphology of an urbanizing watershed using multivariate elasticity. *Adv. Water Resour.* **2015**, *86*, 147–154. [CrossRef]
12. Vaughan, I.P.; Diamond, M.; Gurnell, A.M.; Hall, K.A.; Jenkins, A.; Milner, N.J.; Naylor, L.A.; Sear, D.A.; Woodward, G.; Ormerod, S.J. Integrating ecology with hydromorphology: A priority for river science and management. *Aquat. Conserv. Mar. Freshw. Ecosyst.* **2009**, *19*, 113–125. [CrossRef]
13. Vogel, R.M. Hydromorphology. *J. Water Resour. Plan. Manag.* **2011**, *137*, 147–149. [CrossRef]
14. European Commission. Directive 2000/60/EC of the European Parliament and of the Council of 23 October 2000 establishing a framework for community action in the field of water policy. *Off. J. Eur. Community* **2000**, *60*, 1–72.
15. *EN 14614*; Water Quality—Guidance Standard for Assessing the Hydromorphological Features of Rivers. CEN: Brussels, Belgium, 2004; 21. Available online: http://www.safrass.com/partners_area/BSI%20Hydromorphology.pdf (accessed on 27 November 2021).
16. *EN 15843*; Water Quality—Guidance Standard on Determining the Degree of Modification of River Hydromorphology. CEN: Brussels, Belgium, 2010; 24.
17. Frissell, C.A.; Liss, W.J.; Warren, C.E.; Hurley, M.D. A hierarchical approach to classifying stream habitat features: Viewing streams in a watershed context. *Environ. Manag.* **1986**, *10*, 199–214. [CrossRef]

Article

Use of Pedotransfer Functions in the Rosetta Model to Determine Saturated Hydraulic Conductivity (Ks) of Arable Soils: A Case Study

Łukasz Borek *, Andrzej Bogdał and Tomasz Kowalik

Department of Land Reclamation and Environmental Development, Faculty of Environmental Engineering and Land Surveying, University of Agriculture in Krakow, Al. Mickiewicza 24-28, 30-059 Krakow, Poland; andrzej.bogdal@urk.edu.pl (A.B.); tomasz.kowalik@urk.edu.pl (T.K.)
* Correspondence: lukasz.borek@urk.edu.pl

Abstract: A key parameter for the design of soil drainage and irrigation facilities and for the modelling of surface runoff and erosion phenomena in land-formed areas is the saturated hydraulic conductivity (Ks). There are many methods for determining its value. In situ and laboratory measurements are commonly regarded as the most accurate and direct methods; however, they are costly and time-consuming. Alternatives can be found in the increasingly popular models of pedotransfer functions (PTFs), which can be used for rapid determination of soil hydrophysical parameters. This study presents an analysis of the Ks values obtained from in situ measurements conducted using a double-ring infiltrometer (DRI). The measurements were conducted using a laboratory permeability meter (LPM) and were estimated using five PTFs in the Rosetta program, based on easily accessible input data, i.e., the soil type, content of various grain sizes in %, density, and water content at 2.5 and 4.2 pF, respectively. The degrees of matching between the results from the PTF models and the values obtained from the in situ and laboratory measurements were investigated based on the root-mean-square deviation (*RMSD*), Nash–Sutcliffe efficiency (*NSE*), and determination coefficient (R2). The statistical relationships between the tested variables tested were confirmed using Spearman's rank correlation coefficient (rho). Data analysis showed that in situ measurements of Ks were only significantly correlated with the laboratory tests conducted on intact samples; the values obtained in situ were much higher. The high sensitivity of Ks to biotic and abiotic factors, especially in the upper soil horizons, did not allow for a satisfactory match between the values from the in situ measurements and those obtained from the PTFs. In contrast, the laboratory measurements, showed a significant correlation with the Ks values, as estimated by the models PTF-2 to PTF-5; the best match was found for PTF-2.

Keywords: soil; saturated hydraulic conductivity; pedotransfer function; Rosetta program; irrigation; climate change

Citation: Borek, Ł.; Bogdał, A.; Kowalik, T. Use of Pedotransfer Functions in the Rosetta Model to Determine Saturated Hydraulic Conductivity (Ks) of Arable Soils: A Case Study. *Land* **2021**, *10*, 959. https://doi.org/10.3390/land10090959

Academic Editor: Claude Hammecker

Received: 10 August 2021
Accepted: 8 September 2021
Published: 10 September 2021

Publisher's Note: MDPI stays neutral with regard to jurisdictional claims in published maps and institutional affiliations.

Copyright: © 2021 by the authors. Licensee MDPI, Basel, Switzerland. This article is an open access article distributed under the terms and conditions of the Creative Commons Attribution (CC BY) license (https://creativecommons.org/licenses/by/4.0/).

1. Introduction

Water permeability is a key property of soils, especially with regard to the design of soil irrigation and drainage facilities, modelling of surface runoff and erosion phenomena in land-formed areas, and environmental processes occurring in porous media [1–3]. The study of soil physics, and in particular the determination of the filtration coefficient (Ks) based on direct methods, is a very interesting issue; nevertheless, it is both labour-intensive and costly [4–9].

Climate change in a region or environment entail changes in the method of soil cultivation and plant production. Changes in the interactions between agriculture (i.e., soil compaction) and natural environment (i.e., weather, soil conditions) are a key feature of the transitions which scientists are trying to explain. Therefore, learning the landscape is an important tool in the proper management of water resources in rural areas. Agricultural development depends on the temperature and rainfall distribution during the growing season and on the fertilities and properties of soils, including their abilities to conduct and retain water. The climatic and soil conditions are often unfavourable for the rational and efficient management of environmental resources. Moreover, global climate change, as characterised by an increased frequency of periods of excesses and shortages of water, may hinder the development of modern agriculture, i.e., by reducing the quantity and quality of crop yields. To counteract these adverse phenomena, water engineering and land reclamation facilities are being used in many regions of the world [10,11]. However, to correctly design the spacing for drains, drainage elements, and irrigation ditches and/or the technical parameters of irrigation devices influencing the intensity of the water supply, it is crucial to identify the water permeability of the corresponding soils, including their infiltration and filtration processes [12,13].

The saturated hydraulic conductivity (i.e., Ks) is a key input parameter for modelling the water flow in a soil [14]. Its value can be determined using in situ and laboratory methods, but this often turns out to be a difficult task, owing to the large spatial and temporal variabilities of soil properties [15], and the need to take special care of samples (e.g., to provide an intact structure [16]. Ks is a basic physical property of the soil that affects all soil-plant-water relationships and processes. However, it is one of the most variable soil properties, as it is related to the soil texture and structure, and is influenced by factors such as the land topography, vegetation, land use, and climate [17]. It is widely accepted that in situ measurements provide the most accurate results for the determination of a soil's physical and hydraulic properties.

There are different ways to determine the movements of water in unsaturated (infiltration) and saturated (filtration) zones that are of interest to the scientific community and engineers in various industries. The high cost and effort of in situ measurements and the growing demand for such data have led scientists to seek alternative indirect methods [18,19]. In recent years, pedotransfer functions (PTFs) have become popular and have been used to estimate time-consuming and difficult-to-measure soil properties, such as Ks or soil water retention (pF) values [1,20–22]. Ahuja et al. [23], Rawls et al. [24], Timlin et al. [25], and Suleiman and Ritchie [26] proposed formulas for estimating Ks values based on the effective porosity. Cosby et al. [27], Wösten et al. [28], Saxton and Rawls [29], and Weynants et al. [30] proposed formulas for determining Ks values based on using the textural composition, density, and other physical properties of soils. Currently, studies demonstrating the possibility of using neural networks to estimate Ks values [5,31] are very popular, along with studies aiming to improve existing formulas with appropriate amendments [32]. Puckett et al. [33] proposed a model to predict Ks based on only clay-sized particles. The authors showed that fine sand, sand, and clay percentages were highly correlated with Ks. Jabro [4] developed a model that used the site-average dry bulk density and grain size as predictive variables of Ks. A simple Campbell's model, with soil texture data, can be used [34] or Smettem and Bristow's model from soil clay content using a variety of agricultural topsoil samples [35,36] published a model which were used the relationships between soil texture and soil moisture content at saturation and soil texture and Ks.

Pedotransfer functions, as a five hierarchical models, offered by Rosetta are one of the most applied PTFs, as demonstrated by a number of citations in other research papers more than 1000 times (in Google scholar website). One of the reasons for the Rosetta PTFs popularity is its easy availability and an extensive soil database from North America and Europe containing 2134 soil samples with water retention data and measurements for the saturated hydraulic conductivity (Ks) for 1306 of soils samples and continues to be developed [37]. The majority of work on the performance of PTFs in Rosetta with the aim of determining Ks factor has focused on surface layers of the soil (0–30 cm or 0–50 cm) [13,14,38].

In this work concentrated on the full depth soil pits (150 cm) and including an estimate of Ks for surface layers (0–50 cm) and additionally for a deeper levels of soil (50–150 cm). It was considered that for engineering purposes, such as irrigation or drainage planning, it is advisable to know the Ks parameter.

The aim of this research was to test the possibility of using PTFs in the Rosetta program to determine the Ks values of arable soils based on easily accessible input data, i.e., the soil type, granulometric composition, density, and water content. We hypothesize that the Ks values obtained from PTFs in the Rosetta program for arable soils correspond to the values obtained from (i) direct in situ and (ii) laboratory measurements. To test the hypothesis we examined the arable soils in Central Europe (district of Racibórz, south Poland) in a 150 cm depth soil pits.

2. Materials and Methods

2.1. Description of Study Area

The study area is located in the southwestern part of the Silesian Voivodeship (Poland) in the district of Racibórz (Figure 1). According to Kondracki's [39] division of Poland into physio-geographical regions, the study area is located in the Central European Plain (31), in the macroregion of the Silesian Lowlands (318.5), and on the border of two mesoregions: the Głubczyce Plateau (318.58) and Racibórz Basin (318.59). The main watercourse flowing through the region is the Oder River, along with its left tributary, the Psina River.

Figure 1. Location of the study area.

2.2. Meteorological Conditions

In terms of climate, the study area is considered as one of the warmest areas in this region. In general, Poland has a mostly temperate climate, in transition between an oceanic climate dominating in the north and west of the country, and a continental climate in the south and east. In the multiannual period from 1971–2000, the mean annual air temperature was 8.5 °C, and total precipitation was 616 mm (according to the Institute of Meteorology and Water Management—station in Racibórz). Higher mean temperatures and a decrease in rainfall have been observed over the last decade. The characteristics of the meteorological conditions are listed in Table 1.

Table 1. The characteristics of meteorological conditions of the study area.

Year or Multi-Year	Months of the Growing Season						Period	
	Apr	May	Jun	Jul	Aug	Sep	Apr–Sep	Jan–Dec
	Sum of precipitation totals (mm)							
2010	69	204	112	95	75	78	633	828
2011	27	69	85	131	62	17	391	508
2012	41	35	75	89	69	58	367	586
2013	21	132	110	14	48	99	424	597
2014	27	137	75	58	92	127	516	645
2015	23	55	28	32	13	17	168	280
2016	45	47	47	116	52	26	333	535
2017	68	36	41	69	44	107	365	562
2018	10	42	64	61	55	39	271	441
2019	31	75	33	30	66	98	333	536
	Mean values							
2010–2019	36	83	67	70	68	67	381	552
1971–2000	45	67	79	94	74	56	415	616
	Mean monthly air temperatures (°C)							
2010	9.0	12.5	17.2	20.4	18.7	12.7	15.1	7.9
2011	10.6	13.7	17.8	17.4	19.2	15.4	15.7	9.2
2012	9.9	15.3	17.7	19.9	19.1	14.7	16.1	9.2
2013	9.0	13.8	16.9	19.7	19.1	12.6	15.2	9.0
2014	10.8	13.8	16.3	20.4	17.4	15.6	15.7	10.5
2015	8.8	13.1	16.8	20.9	22.3	15.6	16.3	10.4
2016	9.0	14.7	18.4	19.6	18.2	16.4	16.1	9.8
2017	7.8	14.3	18.8	19.1	20.1	13.8	15.7	9.7
2018	14.0	17.0	18.5	20.2	21.5	16.0	17.9	10.6
2019	10.4	11.9	21.9	19.6	20.9	14.7	16.6	10.8
	Mean values							
2010–2019	9.9	14.0	18.0	19.7	19.7	14.8	16.0	9.7
1971–2000	8.2	13.5	16.1	17.8	17.7	13.6	14.5	8.5
	Values of *HTC* (–)							
2010	2.6	5.3	2.2	1.5	1.3	2.0	Classification of months:	
2011	0.8	1.6	1.6	2.4	1.0	0.4	extremely dry	
2012	1.4	0.7	1.4	1.4	1.2	1.3	very dry	
2013	0.8	3.1	2.2	0.2	0.8	2.6	dry	
2014	0.8	3.2	1.5	0.9	1.7	2.7	relatively dry	
2015	0.9	1.4	0.6	0.5	0.2	0.4	optimal	
2016	1.7	1.0	0.9	1.9	0.9	0.5	moderately humid	
2017	2.9	0.8	0.7	1.2	0.7	2.6	humid	
2018	0.2	0.8	1.2	1.0	0.8	0.8	very humid	
2019	1.0	2.0	0.5	0.5	1.0	2.2	extremely humid	
	Mean values							
2010–2019	1.3	2.0	1.3	1.2	1.0	1.6		

To assess the meteorological conditions of the region where the study was conducted, Selyaninov's hydrothermal coefficient (*HTC*) values were calculated, as a measure of the precipitation efficiency in a given month. The calculation was performed as follows [40]:

$$HTC = \frac{10 \cdot P}{\Sigma t} \; (-) \tag{1}$$

In the above, P is the monthly sum of precipitation (mm), and Σt is the sum of the mean daily air temperature values in a given month (in °C).

Based on the value of the HTC, the months of the growing season (April-September) were classified as follows: $HTC \leq 0.4$—extremely dry; $0.4 < HTC \leq 0.7$—very dry; $0.7 < HTC \leq 1.0$—dry; $1.0 < HTC \leq 1.3$—relatively dry; $1.3 < HTC \leq 1.6$—optimal; $1.6 < HTC \leq 2.0$—moderately humid; $2.0 < HTC \leq 2.5$—humid; $2.5 < HTC \leq 3.0$—very humid; $HTC > 3.0$—extremely humid.

Based on the HTC values calculated for the months of the growing season from the 2010–2019, and as shown in Table 1, it was determined that 58.4% of the months were dry (8.3% extremely dry, 13.3% very dry, 28.5% dry, and 8.3% relatively dry); in 13.3% of the months, the conditions were optimal; and 28.3% of the months were humid (8.3% moderately humid, 6.7% humid, 8.3% very humid, and 5.0% extremely humid). In general, the variable meteorological conditions in the study area, characterised by periods with excesses and shortages of water, indicate the need to use drainage and irrigation facilities for agriculture. To determine their technical parameters, information on the soil permeability is needed. This confirms the advisability of conducting research on the possibility of using mathematical models to determine the soil Ks values, as an alternative to time-consuming in situ or laboratory tests.

2.3. Field Measurement and Soil Sampling

The research was conducted on arable lands during the agricultural season from 2012 to 2015. The field soil tests were conducted at 16 measurement-control points (up to a depth of 150 cm) in Wojnowice (2 points), Bojanów (2 points), Owsiszcze (6 points), Strzybnik (2 points), and Tworków (4 points). Three to five genetic levels were identified for each soil profile.

The soil infiltration was measured in situ, at depths of 10 cm (topsoil) and 35 cm (subsoil), using a double-ring infiltrometer (DRI) method (Figure 2). The DRI comprised an inner ring (9.5 cm diameter) and outer ring (19.5 cm diameter) inserted into the ground at a depth of 10 cm. The DRI was inserted by using a falling weight-type hammer striking on a wooden plank placed uniformly on top of the ring, and without undue disturbance to the soil surface. Each ring of the DRI was filled with a constant head of water level, and the outer ring helped when checking the lateral flow from the inside ring, so as to better estimate the infiltration, reducing losses. The Ks value was estimated when the water flow rate inside the inner ring reached a steady state [1], which in the case of the studied soils, lasted approximately 3–4 h. The infiltration rate was calculated for the respective time intervals ΔT as follows [41]:

$$i = \frac{864 \cdot 4 \cdot V}{3.14 \cdot D_r^2 \cdot \Delta T} \ (\text{m} \cdot \text{day}^{-1}) \tag{2}$$

where V is the volume of water (cm^3) added to the inner ring at time ΔT (s), and D_r is the diameter of the inner ring (cm^2).

For steady infiltration (as a constant value for the soil), the Land and Water Development Division [42] developed infiltration classes, as follows: very slow—<0.024 m·day^{-1}; slow—0.024 ÷ 0.12 m·day^{-1}; moderately slow—0.12 ÷ 0.48 m·day^{-1}; moderate—0.48 ÷ 1.56 m·day^{-1}; moderately rapid—1.56 ÷ 4.20 m·day^{-1}; rapid—4.20 ÷ 5.81 m·day^{-1}; and very rapid—>5.81 m·day^{-1}.

Undisturbed soil samples were taken from each genetic horizon, using cylinders with a volume of 100 cm^3 (three replicates). In addition, approximately 1.0 kg of disturbed soil from each genetic horizon was used to determine the soil texture, and other laboratory analyses were performed.

Figure 2. The double-ring infiltrometer (DRI) (photo. Ł. Borek).

2.4. Laboratory Analysis

The properties of the collected soil samples were determined as follows.

- The soil texture was determined using the Bouyoucose-Casagrande areometric method, based on a measurement of the density of the soil suspension during progressive sedimentation, and a sieving method to fractionate the sand. The contents of the particle size classes (sand, 2.0–0.05 mm; silt, 0.05–0.002 mm; and clay, <0.002 mm) were determined according to the Soil Taxonomy system from the United States Department of Agriculture [43].
- The soil bulk density (*BD*) was determined based on the *gravimetric method, based on* cylinders (100 cm^3) for determining the mass of the dry soil per volume. The weight of each soil core was determined after drying in an oven at 105 °C for approximately 18–24 h. The dry bulk density for each core sample was then calculated as follows [44]:

$$BD = \frac{M_s}{V_t} \; (\text{g·cm}^{-3})$$

(3)

In the above, M_s is the mass of the dry soil weight (g), and V_t is the volume of the total soil sample (cm^3).

- The soil water retention was investigated based on determining the soil suction using ceramic plates in a 5/15 bar pressure plate extractor. The pressure plate equipment used in this study was manufactured by the American Soil Moisture Equipment Corporation. In engineering practice, the soil suction is usually calculated in units of pF, as follows:

$$pF = \log h \; (-)$$

(4)

Here, h is soil suction pressure (in cm H$_2$O).

In the laboratory, the soil water potentials were measured at 300 cm H$_2$O $\approx$ 33 kPa $\approx$ 2.5 pF (representing the field capacity) and 15,000 cm H$_2$O $\approx$ 1500 kPa $\approx$ 4.2 pF (representing the permanent wilting point) [45]. The soil volumetric water content (θ_V) at pF = 2.5 and 4.2 was determined based on the *gravimetric method*. In particular, it was calculated as the ratio of the amount of water in the soil sample to the dry weight of the soil, after drying in an oven at 105 °C for approximately 18–24 h, as follows [44]:

$$\theta_V = \frac{V_{H_2O}}{V_t} \left(\text{cm}^3 \cdot \text{cm}^{-3}\right)$$

(5)

In the above, V_{H2O} is the volume of water in the soil sample (cm^3), and V_t is the volume of the total soil sample (cm^3).

- The Ks was measured under laboratory conditions using a laboratory permeameter (Figure 3) and using the Darcy's law [46] with constant head method (Equation (6)) and falling head method (Equation (7) on undisturbed soil samples for three replications. The constant method can be used with virtually any soil, apart from poorly permeable soils such as clay, whereas the falling-head method is used to measure low-permeability soils, such as f.i. clay or peat samples [47,48]. The soil samples were placed in the laboratory permeameter and then saturated in water for 2–3 days. For the purpose of this study, a permeameter produced by the Eijkelkamp with a closed or open system and 25 holders was used. The Ks was calculated using the Darcy's [46] equation, as follows:

$$V = K \cdot i \cdot A \cdot t \tag{6}$$

Here, explanations below;

Determining Ks using the constant head method was calculated from the following equation:

$$Ks = \frac{V \cdot L}{A \cdot t \cdot h} \tag{7}$$

Here, V is the volume of water flowing through the sample (cm^3); K is the permeability coefficient or 'Ks' (cm·day^{-1} or m·day^{-1}); h is the water level difference inside and outside ringholder or sample cylinder (cm); L is the length of the soil sample (cm); i is the permeability rise gradient or h/L (–); A is the cross-sectional area of the sample (m^2); and t is the time used for flow through of water volume (day). During measuring the following parameters have been determined: L and A—constants, depending on the type of sample ring used; V—volume measured in the burette (1 mL = 1 cm^3); t—length of time lapse; h—calculated with the water levels measured with the water level meter.

Determining Ks using the falling head method was calculated from the following equation:

$$K_s = \frac{a \cdot L}{A \cdot (t_2 - t_1)} \cdot ln\frac{h_1}{h_2} + \frac{x \cdot a \cdot L}{A \cdot \sqrt{(h_1 \cdot h_2)}} \tag{8}$$

Here, Ks is the permeability coefficient (cm·day^{-1} or m·day^{-1}); h is the water level difference inside and outside ringholder or sample cylinder (cm) h_1 and h_2 water level difference inside and outside the ringholder at respectively t_1 (start) and t_2 (end); A is the surface of a cross-section of the sample (cm^2); L is the length of the soil sample (cm); t is the time between beginning and end of the measuring $t_2 - t_1$ (day); a is the cross-section surface of a ringholder or sample cylinder (cm^2) for a sample cylinder applies $A = a$; x is the evaporation factor (literature value): 0.0864 cm·day^{-1} or 0.000864 m·day^{-1}.

The permeability of the soil is also determined by viscosity of the soil solution. Viscosity depends on the temperature. The laboratory water temperature varies from 18 to 22 °C, whereas the average groundwater temperature is 10 °C. Therefore, for certain applications the permeability will have to be corrected for the viscosity of the soil solution (usually water). The Ks for viscosity was corrected using the following equation [49]:

$$K_{10} = K_T \cdot \frac{h_T}{h_{10}} \tag{9}$$

Here, K_{10} is the corrected Ks at 10 °C (cm·day^{-1} or m·day^{-1}); K_T is the Ks at the applied temperature (cm·day^{-1} or m·day^{-1}); h_{10} is the dynamic viscosity of water at 10 °C (Pa·s); h_T is the dynamic viscosity of water at T °C (Pa·s).

The soil classification was established according to the Polish Soil Classification [50], World Reference Base for Soil Resources (FAO and International Union of Soil Sciences (IUSS) Working Group World Reference Base (WRB), [51]), and USDA soil taxonomy (Soil Survey Staff, [43]).

Figure 3. The laboratory permeameter produced by the Eijkelkamp (photo. Ł. Borek).

2.5. Rosetta Description

The Ks values were calculated using the published neural network program Rosetta, with hierarchical PTFs based on five levels of input data [5]. The first level (PTF-1) used soil textural classes based on a lookup table providing parameter means for each USDA textural class. The second level (PTF-2) used the sand, silt, and clay percentages as inputs, and in contrast to PTF-1, provided hydraulic parameters that varied continuously with the texture. The third level (PTF-3) included the predictors used in the level PTF-2, along with the soil dry-bulk density. The fourth level (PTF-4) used PTF-3 and the soil volumetric water content (θ) at a water suction of 33 kPa (2.5 pF). The last level (PTF-5) comprised all of the other parameters, plus the θ value at a water suction of 1500 kPa (4.2 pF) [38,52]. The necessary input data for the Rosetta program (e.g., % soil texture group and bulk density) were determined based on the laboratory analyses of soil samples from the different soil horizons (Tables 2–4).

In the Rosetta program, the relationship between θ and the water suction (h), i.e., the water retention $\theta(h)$, as well as Ks, are described using the well-known Mualem–van Genuchten equations [53] and are given as follows:

$$\theta(h) = \theta_r + \frac{\theta_s - \theta_r}{\left[1 + \left(\alpha|h|^n\right]^m}\tag{10}$$

$$K(h) = K_s \frac{\left\{1 - (\alpha \cdot h)^{n-1}[1 + (\alpha \cdot h)^n)]^m\right\}^2}{\left[1 + (\alpha \cdot h)^n\right]^{m/2}}\tag{11}$$

In the above, $\theta(h)$ is the soil volumetric water content ($cm^3 \cdot cm^{-3}$) at suction h (cm H_2O); θs and θr are the saturated and residual water contents ($cm^3 \cdot cm^{-3}$) at $h = 0$ and 15,000 cm H_2O, respectively; α (>0 in cm^{-1}) is related to the inverse of the air entry suction; n (>1) is a measure of the pore-size distribution, and $m = 1 - 1/n$; and Ks is the saturated hydraulic conductivity ($m \cdot day^{-1}$), as mentioned above.

Table 2. Selected soil physical properties over all depths.

Object Name Profile Number	Depth (cm)	Horizon	Soil Texture Group *	% Fraction			Bulk Density ρ_b (g·cm^{-3})	$\theta_{2.5\,pF}$ (cm^3·cm^{-3})	$\theta_{4.2\,pF}$
				Sand	Silt	Clay			
Wojnowice No. 1	0–28	Ap	SiL	30	60	10	1.64	0.273	0.105
	28–62	Etg	SiL	20	69	11	1.62	0.305	0.085
	62–80	Btg1	SiL	15	68	17	1.66	0.337	0.141
	80–120	Btg2	SL	54	35	11	1.91	0.194	0.108
	120–150	Cg	SiL	21	63	16	1.80	0.297	0.137
Wojnowice No. 2	0–30	Ap	SiL	33	56	11	1.61	0.314	0.122
	30–62	Etg	SiL	18	70	12	1.58	0.330	0.102
	62–94	Btg1	SL	74	17	9	1.86	0.260	0.089
	94–116	Btg2	L	39	49	12	1.84	0.387	0.130
	116–150	Cg	SiL	21	63	16	1.72	0.337	0.132
Strzybnik No. 1	0–27	Ap	SiL	26	61	13	1.54	0.210	0.135
	27–100	Bw	SiL	11	70	19	1.69	0.313	0.150
	100–150	C	SiL	12	72	16	1.73	0.335	0.163
Strzybnik No. 2	0–40	Ap	SiL	25	61	14	1.53	0.302	0.132
	40–90	Bw	SiL	12	69	19	1.59	0.325	0.172
	90–150	C	SiL	12	72	16	1.62	0.336	0.124
Owsiszcze No. 1	0–20	Ap	SiL	25	65	10	1.67	0.334	0.129
	20–38	Etg	SiL	21	68	11	1.64	0.320	0.110
	38–65	Btg	SiL	16	64	20	1.68	0.364	0.177
	65–150	Cg	SiL	16	65	19	1.74	0.342	0.153
Owsiszcze No. 2	0–25	Ap	SiL	25	65	10	1.42	0.312	0.129
	25–43	Etg	SiL	21	68	11	1.58	0.343	0.110
	43–70	Btg	SiL	16	64	20	1.60	0.372	0.177
	70–150	Cg	SiL	16	65	19	1.60	0.370	0.153
Owsiszcze No. 3	0–23	Ap	SiL	26	63	11	1.50	0.328	0.029
	23–36	Etg	SiL	19	69	12	1.65	0.320	0.109
	36–68	Btg	SiL	17	63	20	1.59	0.372	0.181
	68–150	Cg	SiL	16	65	19	1.69	0.338	0.171
Owsiszcze No. 4	0–28	Ap	SiL	26	63	11	1.33	0.340	0.029
	28–41	Etg	SiL	19	69	12	1.56	0.352	0.109
	41–73	Btg	SiL	17	63	20	1.58	0.367	0.181
	73–150	Cg	SiL	16	65	19	1.63	0.344	0.171
Owsiszcze No. 5	0–25	Ap	SiL	27	61	12	1.57	0.352	0.103
	25–42	Etg	SiL	20	68	12	1.60	0.331	0.124
	42–65	Btg	SiL	17	63	20	1.60	0.363	0.168
	65–150	Cg	SiL	15	66	19	1.65	0.339	0.135
Owsiszcze No. 6	0–30	Ap	SiL	27	61	12	1.40	0.351	0.103
	30–47	Etg	SiL	20	68	12	1.58	0.358	0.124
	47–70	Btg	SiL	17	63	20	1.56	0.338	0.168
	70–150	Cg	SiL	15	66	19	1.64	0.344	0.135
Bojanów No. 1	0–25	Ap	SiL	34	53	13	1.80	0.303	0.113
	25–55	Etg	SL	54	30	16	1.96	0.246	0.115
	55–117	Btg	SCL	53	26	21	1.95	0.240	0.194
	117–150	Cg	SL	53	28	19	1.95	0.259	0.166
Bojanów No. 2	0–28	Ap	SiL	25	63	12	1.65	0.316	0.152
	28–63	Etg	SiL	16	68	16	1.69	0.299	0.175
	63–106	Btg	SCL	60	20	20	1.92	0.246	0.156
	106–150	Cg	SL	61	20	19	1.97	0.248	0.138

Table 2. *Cont.*

| Object Name Profile Number | Depth | Horizon | Soil Texture Group * | Physical Characteristics of Soils | | | | | |
| | | | | % Fraction | | | Bulk Density ρ_b | $\theta_{2.5\,pF}$ | $\theta_{4.2\,pF}$ |
	(cm)			Sand	Silt	Clay	(g·cm^{-3})	(cm^3·cm^{-3})	
Tworków No. 1	0–30	Ap	SiCL	18	47	35	1.44	0.427	0.269
	30–46	AC	C	31	26	43	1.12	0.526	0.317
	46–63	OCg	C	16	19	65	1.06	0.609	0.331
	63–94	2O	SCL	57	29	14	1.44	0.676	0.309
	94–150	3G	SiL	19	62	19	1.40	0.488	0.192
Tworków No. 2	0–18	Ap	CL	25	35	40	1.32	0.474	0.247
	18–34	AC	C	18	21	61	1.29	0.516	0.289
	34–52	G	L	32	44	24	1.52	0.448	0.256
	52–85	Cg1	SL	55	33	12	1.75	0.354	0.133
	85–150	2G	SL	76	16	8	1.61	0.345	0.103
Tworków No. 3	0–25	Ap	L	43	35	22	1.65	0.377	0.250
	25–47	A/Bw	L	45	33	22	1.81	0.299	0.150
	47–75	Bw	SL	58	24	18	1.66	0.335	0.158
	75–117	Cg	L	40	43	17	1.68	0.348	0.131
	117–150	2Cg	L	41	45	14	1.68	0.342	0.089
Tworków No. 4	0–30	Ap	CL	22	40	38	1.11	0.566	0.294
	30–42	O	SiCL	17	55	28	1.51	0.439	0.287
	42–71	Bw	SiL	27	53	20	1.56	0.423	0.187
	71–100	Cg	SL	54	33	13	1.65	0.269	0.074
	100–150	2Cg	SL	67	25	8	1.62	0.309	0.065

* soil texture group according to USDA (1999): SL—sandy loam; SCL—sandy clay loam; L—loam; CL—clay loam; SiL—silt loam; SiCL—silty clay loam; C—clay.

Table 3. Basic descriptive statistics for soil physical properties for topsoils and subsoils (n = 32).

| Index Value | Physical Characteristics of Soils | | | | | |
| | % Fraction | | | Bulk Density ρ_b | $\theta_{2.5\,pF}$ | $\theta_{4.2\,pF}$ |
	Sand	Silt	Clay	(g·cm^{-3})	(cm^3·cm^{-3})	
Minimum	11	21	10	1.11	0.210	0.029
Maximum	54	70	61	1.96	0.566	0.317
Mean	25	57	19	1.55	0.350	0.152
Median	25	62	12	1.58	0.329	0.127
SD	9.21	14.77	12.21	0.18	0.079	0.075
CV (%)	36.9	26.1	66.1	11.7	22.5	49.5
n	32	32	32	32	32	32

Table 4. Basic descriptive statistics for soil physical properties over all depths (n = 68).

| Index Value | Physical Characteristics of Soils | | | | | |
| | % Fraction | | | Bulk Density ρ_b | $\theta_{2.5\,pF}$ | $\theta_{4.2\,pF}$ |
	Sand	Silt	Clay	(g·cm^{-3})	(cm^3·cm^{-3})	
Minimum	11	16	8	1.11	0.194	0.029
Maximum	76	72	65	1.97	0.676	0.331
Mean	30	52	18	1.60	0.351	0.155
Median	24	62	16	1.62	0.338	0.138
SD	16.8	17.6	10.5	0.23	0.086	0.065
CV (%)	57.0	33.9	57.2	14.6	24.4	41.8
n	68	68	68	68	68	68

2.6. Statistical Analysis and Model Performance Evaluation

The data set comprised the analytical results from the soil samples collected from the 16 soil pits. For the statistical analysis, the procedures provided by the program *Statistica PL Version 12.5* were used, with a 5% significance level. The minimum and maximum values were determined for each physical parameter of the soil, and the arithmetic mean, median, standard deviation (*SD*), and coefficient of variation (*CV*) were computed. Moreover, a Spearman correlation test was conducted for the dataset. The correlation strength and direction of the relationship between two variables were determined based on Spearman's rank correlation coefficient (*rho*), which takes values from −1 to 1. A positive sign of the coefficient indicated the existence of a positive correlation, and vice versa. The closer the values were to −1 and 1, the stronger the correlation. When *rho* = 0, there was no correlation between the examined variables [54]. This statistical method was chosen after finding that most data were not normally distributed (Shapiro-Wilk test). All of data are presented in the tables and graphs, i.e., to produce a visual image that is helpful in interpreting the results.

Based on the available soil data in this study, five widely used PTFs were selected to assess their respective performances in estimating Ks, via comparison with the measured Ks values (via the DRI) for the soil cores collected in the field (using a laboratory permeameter). The calculated values of Ks obtained using Rosetta were compared to the corresponding measured values and evaluated using two statistical parameters: root-mean-square deviation (*RMSD*) and Nash–Sutcliffe efficiency (*NSE*). The *RMSD* gives the mean difference between the measured and calculated values of Ks, and is calculated as follows [38,55]:

$$RMSD = \sqrt{\frac{1}{n}\sum_{i=1}^{n}(m_i - p_i)^2} \tag{12}$$

where: m_i is the measured value of Ks, and p_i is the corresponding (predicted) value of Ks as obtained using the Rosetta program.

The *RMSD* values are always non-negative and should be as low as possible; *RMSD* = 0 indicates a perfect fit of the model to the measurement data.

The *NSE* compares measured and predicted values (here, for Ks), and is given as follows [56]:

$$NSE = 1 - \left[\frac{\sum_{i=1}^{n}(m_i - p_i)^2}{\sum_{i=1}^{n}(m_i - \overline{m}_i)^2}\right] \tag{13}$$

In the above, n is the number of observations, m_i and p_i are the measured and predicted values of Ks, respectively, and $\overline{m}_i$ is the mean of the measured Ks values.

The *NSE* values range from −∞ to 1. A value of *NSE* = 1 corresponded to a perfect match of calculated values to the measured values of Ks. The 'efficiency *NSE*' = 0 indicated that the calculations of Ks using the Rosetta program were as accurate as the mean of the measured data; in contrast, an efficiency *NSE* < 0 occurred when the measured mean was a better predictor than the model or, in other words, when the residual variance, as described by the numerator in the equation above, was larger than the data variance described by the denominator.

3. Results and Discussion

According to the PTG [50], FAO and IUSS Working Group WRB [52], and USDA soil taxonomy [43], the examined soils were classified as follows.

- Order 3: Brown forest soils (brown earths, PTG—Polish: *Gleby brunatnoziemne*; WRB: Cambisols; USDA: Inceptisols—Udepts), Type 3.1. 'Euthrophic brown soils' (PTG—Polish: Gleby brunatne eutroficzne; WRB: Haplic Cambisol, Haplic, Stagnic, Endogleyic, or Vertic Cambisol (Eutric); USDA: Typic or Humic or Aquic or Oxyaquic or Vertic Eutrudepts)—occurred in Strzybnik.

- Order 5: Brown forest podzolic soils (Soil lessivé) (PTG—Polish: *Gleby płowoziemne*; WRB: Luvisols, Albeluvisols; USDA: Alfisols—Aqualfs, Udalfs), Type 5.2. 'Streak brown forest podzolic soils' (PTG—Polish: Gleby płowe zaciekowe; WRB: Haplic, Stagnic, Gleyic, Cambic Albeluvisol; USDA: Typic, Arenic, Aquic, Oxyaquic, Haplic, Glossaqiuc, or Vertic Glossudalfs)—occurred in Wojnowice, Bojanów and Owsiszcze.
- Order 7: Chernozemic soils (PTG—Polish: *Gleby czarnoziemne*; WRB: Chernozems, Phaeozems; USDA: Mollisols—Aquolls, Udolls), Type 7.4. 'Chernoziemic fluvisols' (PTG—Polish: Mady czarnoziemne; WRB: Mollic Fluvisol, Endofluvic Phaeozem; USDA: Fluvaquentic Endoaquolls)—occurred in Tworków.

As shown in Table 2 and Figure 4, the examined soils are heterogeneous in terms of their textures. Seven textural classes are observed in the examined soil profiles: silt loam (SiL; $n = 42$), sandy loam (SL; $n = 10$), loam (L; $n = 6$), clay (C; $n = 3$), sandy clay loam (SCL; $n = 3$), clay loam (CL; $n = 2$), and silty clay loam (SiCL; $n = 2$). The silt loam is the predominant soil texture. The *SD* values (Tables 3 and 4) for most soil properties are large, indicating moderate to strong variability; this is not uncommon for soils [12,57,58]. Typically, the *CV* values are < 45%, indicating the mean variability of the measurement data. The highest *CV* values are observed for sand (57.0%) and clay (57.2%). The sand content in the soil profiles is between 11% and 76%, the silt content is between 16% and 72%, and the clay content is between 8% and 65% (Tables 2 and 4). The high swelling-clay content could be a cause of the poor water flow through the samples, as well as a consequence of abnormal Ks results [59].

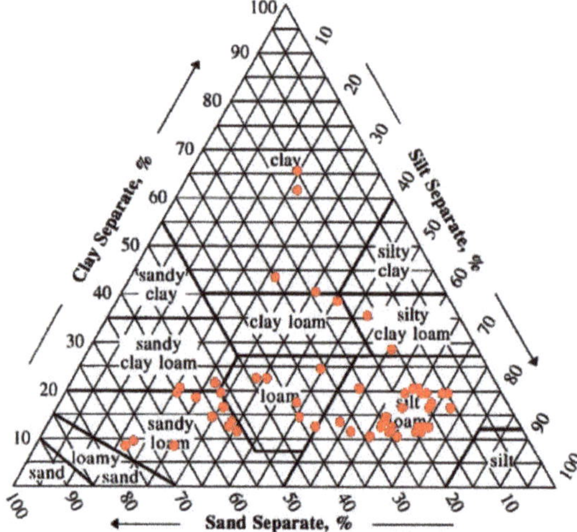

Figure 4. Textural composition of soils investigated.

With regard to the surface soil horizons ($n = 32$), the mean soil bulk density (ρ_b) is 1.55 g·cm^{-3}, whereas that for overall depths is 1.60 g·cm^{-3} (Tables 3 and 4). The soil volumetric water contents at the matric potentials of 33 kPa (2.5 pF) and 1500 kPa (4.2 pF) in both cases have similar mean values (0.350/0.351 cm^3·cm^{-3} and 0.152/0.155 cm^3·cm^{-3}, respectively).

Higher values of Ks are found in the upper horizons of the tested soils, owing to the significant macroporosity caused by plant roots, reconsolidation [60], and annual loosening from agrotechnical works and alternating soil freezing and thawing [61]. In addition, zoogenic channels (especially earthworm channels) and plant residues increase soil permeability [62]. For these reasons, the samples taken from arable and sub-arable horizons are less dense than those from lower horizons; the latter are also subject to self-compacting (Tables 2–4).

Table 5 presents descriptive statistics concerning the values of Ks for the topsoils and subsoils (*n* = 32), as obtained from in situ and laboratory measurements, and as estimated with the use of the PTF models in the Rosetta program (variant 1). In contrast, Table 6 presents statistical measures considering the Ks values of all of the soil horizons (*n* = 68), but excludes the in situ tests not conducted on larger and deeper horizons, for practical reasons (variant 2).

Table 5. Basic descriptive statistics for soil hydraulic conductivity (Ks) for topsoils and subsoils.

Index Value	Ks (m·day^{-1})						
	Field Measured	Laboratory Measured	PTF-1	PTF-2	PTF-3	PTF-4	PTF-5
Minimum	0.01	0.01	0.08	0.07	0.62	0.24	0.27
Maximum	4.67	1.18	0.38	0.40	1.82	1.90	1.91
Mean	0.85	0.21	0.17	0.25	1.41	0.45	0.45
Median	0.40	0.11	0.18	0.28	1.60	0.33	0.31
SD	1.09	0.29	0.05	0.10	0.37	0.34	0.34
CV (%)	128.8	138.7	29.2	40.5	26.1	75.7	75.2
n	32	32	32	32	32	32	32

Table 6. Basic descriptive statistics for soil hydraulic conductivity (Ks) over all depths.

Index Value	Ks (m·day^{-1})					
	Laboratory Measured	PTF-1	PTF-2	PTF-3	PTF-4	PTF-5
Minimum	0.01	0.08	0.07	0.55	0.24	0.27
Maximum	3.48	0.38	0.66	2.71	4.11	3.93
Mean	0.21	0.20	0.23	1.37	0.64	0.64
Median	0.04	0.18	0.19	1.32	0.32	0.31
SD	0.49	0.08	0.11	0.40	0.76	0.72
CV (%)	230.7	41.8	48.0	28.7	119.2	111.8
n	68	68	68	68	68	68

In variant 1, the characteristic values of Ks (maximum, mean, and median values) as obtained from the in situ measurements are approximately four times higher than the Ks values obtained from the laboratory measurements. Such large discrepancies between the results obtained from both direct methods result from the manner of conducting the measurements; in particular, small-volume soil samples, which do not allow for the inclusion of all of the factors influencing the soil permeability, were taken for the laboratory tests. The mean Ks value obtained from the in situ measurements, i.e., 0.85 m·day^{-1}, is four times higher than those estimated by PTF-1 and PTF-2, almost half lower than that obtained from PTF-3, and twice as high as the values obtained from PTF-4 and PTF-5. In addition, the median value from the in situ tests (0.40 m·day^{-1}) is from 1.2 to 2.2 times higher than the values obtained from the four PTF models—the exception is model PTF-3, whose median value is four times higher than that from the in situ tests. The maximum and mean values of the Ks values obtained from laboratory tests, compared to those from the in situ method, are usually more similar to those obtained from the PTF models (Table 5). The mean Ks values obtained from the in situ measurements and PTF-3 indicate the mean class, whereas the results obtained from laboratory measurements and the other four PTF models indicate a moderately slow basic infiltration class [42].

Most of the subsurface horizons are characterised by low Ks values, usually not exceeding 0.50 m·day^{-1}. In variant 2, covering the analysis of all soil horizons, the mean value of Ks from the laboratory measurements (0.21 m·day^{-1}) is very close to the results obtained from the PTF-1 and PTF-2 models, almost six times lower than the Ks value from PTF-3, and three times lower than the values estimated by PTF-4 and PTF-5 (Table 6). The mean Ks values obtained from the laboratory measurements and PTF-1, PTF-2, and PTF-3 indicate the moderate class, whereas the results obtained from the other three PTF models indicate a moderately slow basic infiltration class [42].

Regardless of the amount of data taken for analysis ($n = 32$ or 68), the values of Ks are usually characterised by high or very high random variability ($CV = 26.1$–230.7%), not only in comparison to the other soil hydrophysical properties, but also compared to the methods used to determine Ks, i.e., field, laboratory, and PTF tests (Tables 5 and 6, Figure 5). This regularity has been described in various other studies, for example, Rezaei et al. [63]. As reported by Merdun et al. [64] and García-Gutiérrez et al. [65], the high spatial variability between the soil horizons with regard to Ks are the result of the soil heterogeneity, which hampers the prediction of Ks using mathematical models.

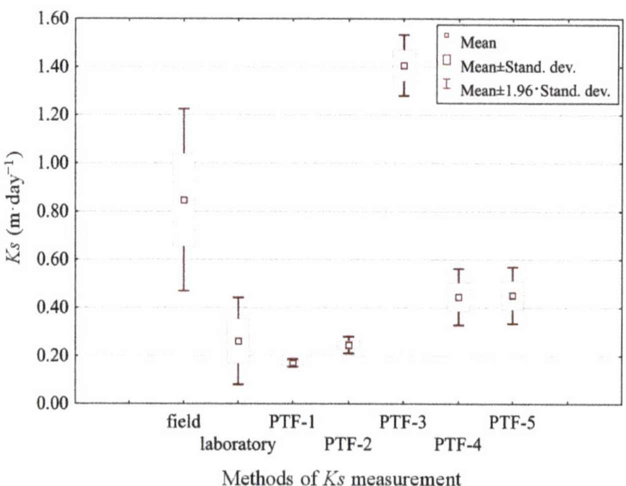

Figure 5. Box plots showing the performance of Ks to different measurement methods.

The Ks values obtained from the laboratory measurements are statistically significant ($p < 0.05$), negatively correlated with the clay fraction and density, and positively correlated with the sand fraction (Table 7). Similar observations were noted by Zhao et al. [13], who conducted studies on the Loess Plateau of China. The interdependence of Ks values and soil textures were well documented by Hillel [66], García-Gutiérrez et al. [65], and other researchers [3]. In the case of in situ measurements, the soil hydraulic conductivity is only significantly positively correlated with the sand fraction ($rho = 0.29$). The PTF-1 model is significantly positively correlated with the soil density, but negatively correlated with the clay fraction and soil volumetric water content (θ) at water suctions of 33 and 1500 kPa (2.5 pF and 4.2 pF). The PTF-2 and PTF-3 models show a significantly positive correlation with the sand fraction and negative correlations with the clay fraction and soil volumetric water content (θ) at 2.5 and 4.2 pF. The PTF-4 and PTF-5 models are significantly positively correlated with the sand fraction, and are negatively correlated with the dust fraction (Table 7).

Table 7. Spearman's rank correlations coefficients (*rho*) between measured and predicted saturated hydraulic conductivity (K_s) and other selected soil physical properties.

Variables	% Sand	% Silt	% Clay	Bulk Density	$\theta_{2,5\,pF}$	$\theta_{4,2\,pF}$	K_s—Field	K_s—Laboratory	K_s—PTF−1	K_s—PTF−2	K_s—PTF−3	K_s—PTF−4	K_s—PTF−5
% Sand	1.00												
% Silt	−0.77	1.00											
% Clay	−0.31	−0.22	1.00										
Bult density	0.23	−0.06	−0.15	1.00									
$\theta_{2,5\,pF}$	−0.29	−0.04	0.47	−0.61	1.00								
$\theta_{4,2\,pF}$	−0.30	−0.19	0.87	−0.23	0.53	1.00							
K_s—Field	0.29	−0.08	−0.21	−0.24	0.06	−0.04	1.00						
K_s—Laboratory	0.29	−0.18	−0.36	−0.30	−0.17	−0.20	0.54	1.00					
K_s—PTF−1	0.15	0.07	−0.54	0.35	−0.50	−0.50	−0.13	0.12	1.00				
K_s—PTF−2	0.34	0.09	−0.93	0.06	−0.43	−0.78	0.15	0.40	0.53	1.00			
K_s—PTF−3	0.37	0.15	−0.96	0.12	−0.47	−0.85	0.21	0.34	0.55	0.95	1.00		
K_s—PTF−4	0.81	−0.75	−0.17	0.17	−0.22	−0.20	0.21	0.35	0.03	0.28	0.20	1.00	
K_s—PTF−5	0.84	−0.90	0.00	0.04	−0.06	−0.04	0.12	0.25	−0.04	0.13	0.06	0.90	1.00

Red color means, that the determined correlation coefficient is significant (for the significance level $\alpha = 0.05$).

No significant correlations were detected between the Ks values obtained from the in situ measurements and the PTF models. However, with the exception of the PTF-1 model, the Ks values from the laboratory tests as estimated by the four PTF models are positively correlated (Table 7). For these reasons, only the linear correlations between the values obtained from the laboratory tests and PTF models are presented in Figure 6a–e.

In general, the matching of the Ks results obtained from the direct in situ measurements with the results estimated by the Rosetta PTFs is poorer than that obtained from the laboratory measurements and estimated by the Rosetta PTFs. This is confirmed by the *RMSD* values, which range from 1.40 to 1.64 in the first case, whereas in the second case, they range from 0.19 to 1.58. In addition, the values of the coefficient of determination (R^2), which indicates the extent to which the model explains the gathered measurement data, are usually much higher when comparing the data obtained from the laboratory tests and that from the PTF models (Table 8).

Table 8. Statistical indices for estimation of saturated hydraulic conductivity (Ks).

Model	Field Measured Ks ($m \cdot day^{-1}$)				Laboratory Measured Ks ($m \cdot day^{-1}$)			
	n	*RMSD*	*NSE*	R^2	n	*RMSD*	*NSE*	R^2
Rosetta—PTF-1	32	1.63	−0.42	0.02	68	0.23	0.04	0.01
Rosetta—PTF-2	32	1.54	−0.34	0.02	68	0.19	0.20	0.16
Rosetta—PTF-3	32	1.64	−0.42	0.04	68	1.58	−6.34	0.12
Rosetta—PTF-4	32	1.40	−0.22	0.04	68	0.65	−2.03	0.12
Rosetta—PTF-5	32	1.40	−0.22	0.01	68	0.61	−1.83	0.06
Laboratory measured	32	1.53	−0.33	0.29				

The *RMSD* values of the Ks results obtained from the laboratory measurements and estimated with PTFs reveal the best match for the case of the PTF-2 model. The *NSE* and R^2 values are the best in this case (Table 8), indicating that the granulometric composition (% of sand, silt, and clay) in the PTF-2 model has the greatest influence on the water permeability of the soils.

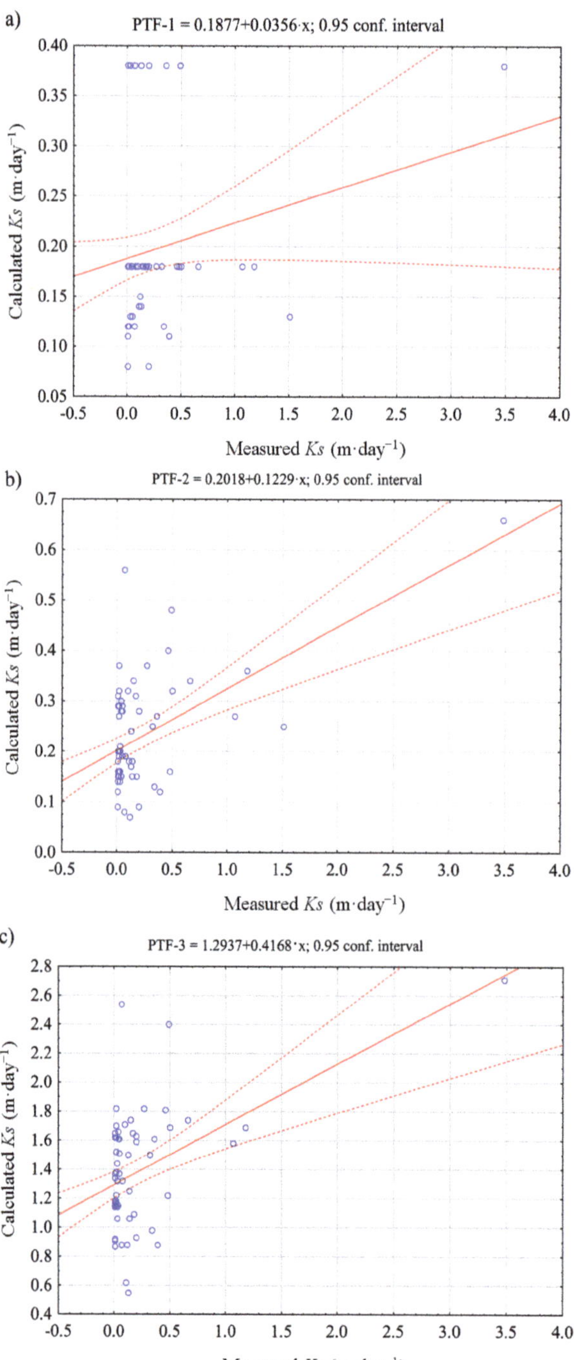

Figure 6. *Cont.*

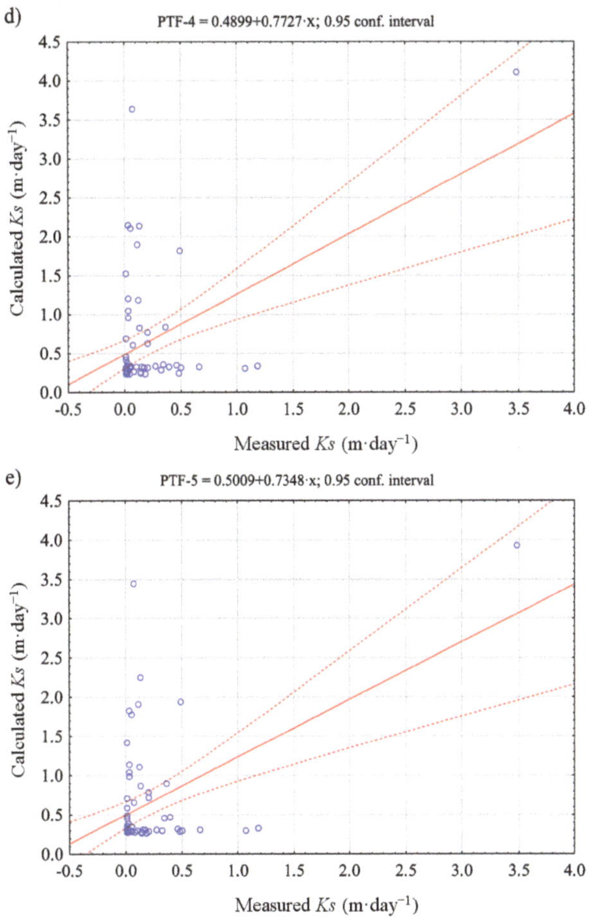

Figure 6. The laboratory measured and predicted saturated hydraulic conductivity Ks for the five pedotransfer function models invistigated (PTF).

4. Conclusions

The search for alternatives to labour-intensive and cost-intensive in situ tests and laboratory methods for determining the value of Ks, at a time of global climate change, seems to be a priority. The increased frequency of periods with excesses and scarcities of water will require the greater use of soil irrigation and drainage facilities, and the modelling of surface runoff and erosion phenomena in land-formed areas. Correspondingly, knowledge regarding soil permeability is required for all technical and non-technical activities.

In general, when using a descriptive comparative analysis, it is difficult to show a significant relationship between Ks values from in situ tests and the diverse hydrophysical properties of arable soils, which are subject to cyclic loosening, reconsolidation, and/or freezing and thawing. The high sensitivity of this physical parameter to biotic and abiotic factors, especially in the soil upper horizons, hampered a satisfactory adjustment to the Ks values obtained from the PTF models. Therefore, in situ tests, although difficult and time-consuming, should continue to be conducted to obtain a reliable measurement database that is free of errors resulting from the soil type variability, and that considers the influences of fauna and flora on the physical properties of soil. This is of particular importance for the modelling of surface runoff and erosion processes and design of surface irrigation systems (e.g., sprinklers), where only 40–50 cm of the upper horizon of the soil is wetted, and the technical parameters of the corresponding drainage facilities are highly dependent on soil permeability.

The in situ measurements of Ks are significantly correlated only with the laboratory tests; nevertheless, the values obtained in situ are much higher, and are influenced by a number of factors, including those of a methodological nature (e.g., the larger diameter of the measuring cylinder). In contrast, the laboratory measurements show a significant correlation with the Ks values estimated using the models PTF-2 to PTF-5. The best match is found (based on the *RMSD*, *NSE*, and R^2 values) when using the PTF-2 model in the Rosetta program. This means that the most reliable Ks values can be obtained from the percentages of the sand, dust, and clay fractions. Therefore, there is a rational alternative, in the form of the PTF, for the costly and labour-intensive determinations of Ks conducted in the laboratory. The use of the PTF-2 model allows for the determination of the hydraulic conductivity of deeper soil horizons, where the variability of the hydrophysical properties is lower and there are large methodological limitations in conducting field work or difficulties in taking intact structural samples for laboratory tests. This type of solution can be used to collect data for designing subsurface drainage and irrigation facilities, some of the technical parameters of which are determined by the Ks values.

In conclusion, the research results confirmed the remarkable difference between the performance of PTFs in the Rosetta and the Ks results obtained from field or laboratory measurements. The study thus provided evidence that the Ks values obtained from local dataset give better results performances in predicting the soil saturated hydraulic conductivity, compared to PTFs in the Rosetta derived from large datasets, which in turn confirm the limitation of applying PTFs developed from one region to other regions.

Author Contributions: Conceptualization, Ł.B., A.B. and T.K.; methodology, Ł.B., A.B. and T.K.; software, Ł.B.; validation, Ł.B.; formal analysis, Ł.B.; investigation, Ł.B., A.B. and T.K.; resources, Ł.B., A.B. and T.K.; data curation, Ł.B., A.B. and T.K.; writing—original draft preparation, Ł.B., A.B. and T.K.; writing—review and editing, Ł.B. and T.K.; visualization, Ł.B.; supervision, T.K.; project administration, Ł.B.; funding acquisition, Ł.B., A.B. and T.K. All authors have read and agreed to the published version of the manuscript.

Funding: This research was funded by the Department of Land Reclamation and Environmental Development, Faculty of Environmental Engineering and Land Surveying, University of Agriculture in Krakow.

Institutional Review Board Statement: Not applicable.

Informed Consent Statement: Not applicable.

Data Availability Statement: Not applicable.

Acknowledgments: The authors would like to acknowledge the authorities of the Agro-Industrial Company 'AGROMAX' in Racibórz for providing agricultural areas and agricultural equipment for the field study.

Conflicts of Interest: The authors declare no conflict of interest.

References

1. Islam, A.; Mailapalli, D.R.; Behera, A. Evaluation of Saturated Hydraulic Conductivity Methods for Different Land Uses. *Indian J. Ecol.* **2017**, *44*, 456–466.
2. Kanso, T.; Tedolidi, D.; Gromaire, M.-C.; Rameier, D.; Saad, M.; Chebbo, G. Horizontal and vertical variability of soil hydraulic properties in roadside sustainable drainage systems (SuDS)—Nature and Implications for Hydrological Performance Evaluation. *Water* **2018**, *10*, 987. [CrossRef]
3. Ottoni, M.V.; Ottoni Filho, T.B.; Lopes-Assad, M.L.R.; Rotunno Filho, O.C. Pedotransfer functions for saturated hydraulic conductivity using a database with temperate and tropical climate soils. *J. Hydrol.* **2019**, *575*, 1345–1358. [CrossRef]
4. Jabro, J.D. Estimation of saturated hydraulic conductivity of soils from particle-size distribution and bulk-density data. *Trans. ASAE* **1992**, *35*, 557–560. [CrossRef]

5. Schaap, M.G.; Leij, F.J.; van Genuchten, M.T. ROSETTA: A computer program for estimating soil hydraulic parameters with hierarchical pedotransfer functions. *J. Hydrol.* **2001**, *251*, 163–176. [CrossRef]

6. Li, Y.; Chen, D.; White, R.E.; Zhu, A.; Zhang, J. Estimating soil hydraulic properties of Fengqiu County soils in the North China plain using pedo-transfer functions. *Geoderma* **2007**, *138*, 261–271. [CrossRef]

7. Jarvis, N.; Koestel, J.; Messing, I.; Moeys, J.; Lindahl, A. Influence of soil, land use and climatic factors on the hydraulic conductivity of soil. *Hydrol. Earth Syst. Sci.* **2013**, *17*, 5185–5195. [CrossRef]

8. Papanicolaou, A.N.; Elhakeem, M.; Wilson, C.G.; Burras, C.L.; West, L.T.; Lin, H.; Clark, B.; Oneal, B.E. Spatial variability of saturated hydraulic conductivity at the hillslope scale: Understanding the role of land management and erosional effect. *Geoderma* **2015**, *243–244*, 58–68. [CrossRef]

9. Kruk, E. Impact of the Technological Path on Some Soil Properties on Loess Slope. *J. Ecol. Eng.* **2019**, *20*, 169–176. [CrossRef]

10. Borek, Ł.; Bogdał, A.; Ostrowski, K. The effect of subsoiling on change of compaction and water permealility of silt loam. *Annu. Set Environ. Prot.* **2018**, *20*, 538–557. Available online: http://towarzystwo.ros.edu.pl/images/roczniki/2018/030_ROS_V20_R2 018.pdf, (accessed on 5 December 2020).

11. Lauffenburger, Z.H.; Gurdak, J.J.; Hobza, C.; Woodward, D.; Wolf, C. Irrigated agriculture and future climate change effects on groundwater recharge, northern High Plains aquifer, USA. *Agric. Water Manag.* **2018**, *204*, 69–80. [CrossRef]

12. Gamie, R.; De Smedt, F. Experimental and statistical study of saturated hydraulic conductivity and relations with other soil properties of a desert soil. *Eur. J. Soil Sci.* **2018**, *69*, 256–264. [CrossRef]

13. Zhao, C.; Shao, M.; Jia, X.; Nasir, M.; Zhang, C. Using pedotransfer functions to estimate soil hydraulic conductivity in the Loess Plateau of China. *CATENA* **2016**, *143*, 1–6. [CrossRef]

14. Jačka, L.; Pavlásek, J.; Kuráž, V.; Pech, P. A comparison of three measuring methods for estimating the saturated hydraulic conductivity in the shallow subsurface layer of mountain podzols. *Geoderma* **2014**, *219–220*, 82–88. [CrossRef]

15. Twarakavi, N.K.C.; Simunek, J.J.; Schaap, M.G. Development of pedotransfer functions for estimation of soil hydraulic parameters using support vector machines. *Soil Sci. Soc. Am. J.* **2009**, *73*, 1443–1452. [CrossRef]

16. Bouma, J. Using soil survey data for quantitative land evaluation. *Adv. Soil Sci.* **1989**, *9*, 177–213. [CrossRef]

17. Deb, S.K.; Shukla, M.K. Variability of hydraulic conductivity due to multiple factors. *Am. J. Environ. Sci.* **2012**, *8*, 489–502. [CrossRef]

18. Gunarathna, M.; Sakai, K.; Nakandakari, T.; Momii, K.; Kumari, M. Machine learning approaches to develop pedotransfer functions for tropical Sri Lankan soils. *Water* **2019**, *11*, 1940. [CrossRef]

19. Givi, J.; Prasher, S.O.; Patel, R.M. Evaluation of pedotransfer functions in predicting the soil water contents at field capacity and wilting point. *Agric. Water Manag.* **2004**, *70*, 83–96. [CrossRef]

20. Borek, Ł.; Bogdał, A. Soil wetar retention of the Odra River alluvial soils (Poland): Estimating parameters by RETC model and laboratory measurements. *Appl. Ecol. Environ. Res.* **2018**, *16*, 4681–4699. Available online: http://www.aloki.hu/pdf/1604_46814 699.pdf (accessed on 11 December 2020). [CrossRef]

21. Wang, J.-P.; Zhuang, P.-Z.; Luan, J.-Y.; Liu, T.-H.; Tan, Y.-R.; Zhang, J. Estimation of unsaturated hydraulic conductivity of granular soils from particle size parameters. *Water* **2019**, *11*, 1826. [CrossRef]

22. Zhang, Y.; Schaap, M.G. Estimation of saturated hydraulic conductivity with pedotransfer functions: A review. *J. Hydrol.* **2019**, *575*, 1011–1030. [CrossRef]

23. Ahuja, L.R.; Cassel, D.K.; Bruce, R.R.; Barnes, B.B. Evaluation of spatial distribution of hydraulic conductivity using effective porosity data. *Soil Sci.* **1989**, *148*, 404–411. [CrossRef]

24. Rawls, W.J.D.; Gimenez, D.; Grossman, R. Use of soil texture, bulk density and slope of the water retention curve to predict saturated hydraulic conductivity. *Trans. ASAE* **1998**, *41*, 983–988. [CrossRef]

25. Timlin, D.J.; Ahuja, L.R.; Pachepsky, Y.; Williams, R.D.; Gimenez, D.; Rawls, W. Use of Brooks-Corey parameters to improve estimates of saturated conductivity from effective porosity. *Soil Sci. Soc. Am. J.* **1999**, *63*, 1086–1092. [CrossRef]

26. Suleiman, A.A.; Ritchie, J.T. Estimating saturated hydraulic conductivity from soil porosity. *Trans. ASAE* **2001**, *44*, 235–339. [CrossRef]

27. Cosby, B.J.; Hornberger, G.M.; Clapp, R.B.; Ginn, T.R. A statistical exploration of the relationships of soil moisture characteristics to the physical properties of soils. *Water Resour. Res.* **1984**, *20*, 682–690. [CrossRef]

28. Wösten, J.H.M.; Lilly, A.; Nemes, A.; Le Bas, C. Development and use of a database of hydraulic properties of European soils. *Geoderma* **1999**, *90*, 169–185. [CrossRef]

29. Saxton, K.E.; Rawls, W.J. Soil water characteristic estimates by texture and organic matter for hydrologic solutions. *Soil Sci. Soc. Am. J.* **2006**, *70*, 1569–1578. [CrossRef]

30. Weynants, M.; Vereecken, H.; Javaux, M. Revisiting vereecken pedotransfer functions: Introducing a closed-form hydraulic model. *Vadose Zone J.* **2009**, *8*, 86–95. [CrossRef]

31. Zhang, Y.; Schaap, M.G. Weighted recalibration of the rosetta pedotransfer model with improved estimates of hydraulic parameter distributions and summary statistics (Rosetta 3). *J. Hydrol.* **2017**, *547*, 39–53. [CrossRef]

32. Araya, S.N.; Ghezzehei, T.A. Using machine learning for prediction of saturated hydraulic conductivity and its sensitivity to soil structural perturbations. *Water Resour. Res.* **2019**, *55*, 5715–5737. [CrossRef]

33. Puckett, W.E.; Dane, J.H.; Hajek, B. Physical and Mineralogical Data to Determine Soil Hydraulic Properties†. *Soil Sci. Soc. Am. J.* **1985**, *49*, 831–836. [CrossRef]

34. Campbell, G.S. *Soil Physics with BASIC: Transport Models for Soil-Plant Systems*; Elsevier Scientific: New York, NY, USA, 1985.
35. Smettem, K.R.J.; Bristow, K.L. Obtaining soil hydraulic properties for water balance and leaching models from survey data. 2. Hydraulic conductivity. *Aust. J. Agric. Res.* **1999**, *50*, 1259–1262. [CrossRef]
36. Saxton, K.E.; Rawls, W.J.; Romberger, J.S.; Papendick, R. Estimating general soil-water characteristics from texture. *Soil Sci. Soc. Am. J.* **1986**, *5*, 1031–1036. [CrossRef]
37. Zhang, Y.; Schaap, M.G.; Zha, Y. A highresolution global map of soil hydraulic properties produced by a hierarchical parameterization of a physically based water retention model. *Water Resour. Res.* **2018**, *54*, 9774–9790. [CrossRef]
38. Alvarez-Acosta, C.; Lascano, R.J.; Stroosnijder, L. Test of the ROSETTA pedotransfer function for saturated hydraulic conductivity. *Open J. Soil Sci.* **2012**, *2*, 203–212. [CrossRef]
39. Kondracki, J. *Regional Geography of Poland*; PWN: Warszawa, Poland, 2011; 440p, ISBN 9788301160227. (In Polish)
40. Skowera, B. Changes of hydrothermal conditions in the Polish area (1971–2010). *Fragm. Agron.* **2014**, *31*, 74–87. (In Polish). Available online: https://pta.up.poznan.pl/pdf/2014/FA%2031(2)_2014_skowera_nowa_wersja.pdf (accessed on 3 December 2020).
41. Gulliver, J.S.; Anderson, J.L. (Eds.) *Assessment of Stormwater Best Management Practices*; Report of Stormwater Management Practice Assessment Project; University of Minnesota: Minneapolis, MN, USA, 2008; 599p.
42. FAO. *Land and Water Development Division*; FAO: Rome, Italy, 1971.
43. Soil Survey Staff. *Soil Taxonomy: A Basic System of Soil Classification for Making and Interpreting Soil Surveys*, 2nd ed.; Handbook; United States Department of Agriculture (USDA), Natural Resources Conservation Service: Washington, DC, USA, 1999; 886p. Available online: https://www.nrcs.usda.gov/Internet/FSE_DOCUMENTS/nrcs142p2_051232.pdf (accessed on 13 December 2020).
44. Phogat, V.K.; Tomar, V.S.; Dahyia, R. Soil Physical Properties (Chapter 6). In *Soil Science: An Introduction*, 1st ed.; Rattan, R.K., Katyal, J.C., Dwivedi, B.S., Sarkar, A.K., Bhattachatyya, T., Tarafdar, J.C., Kukal, S.S., Eds.; Indian Society of Soil Science: New Delhi, India, 2015; pp. 135–171.
45. Mocek, A.; Drzymała, S. *Genesis, Analysis and Classification of Soils*; Wyd. UP: Poznań, Poland, 2010; 418p. (In Polish)
46. Darcy, H. *Les Fontaines Publiques de la Ville de Dijon [The Public Fountains of the City of Dijon]*; Dalmont: Paris, France, 1856. (In France)
47. Klute, A.; Dinauer, R.C.; Page, A.L.; Miller, R.H.; Keeney, D.R. *Methods of Soil Analysis. Part 1. Physical a Mineralogical Methods*; SSSA Book Series No 5; Soil Science Society of America: Madison, WI, USA, 1986; p. 1358. ISBN 978-0891188117.
48. Eijkelkamp. Laboratory Permeameter. Operating Instructions. 2017. Available online: https://en.eijkelkamp.com/products/laboratory-equipment/permeameter-closed-system-25-holders-73766.html;https://en.eijkelkamp.com/products/laboratory-equipment/set-for-pf-determination-with-ceramic-plates.html (accessed on 13 December 2020).
49. Wit, K.E. *Meting van de Doorlatendheid in Ongeroerde Monsters*; Rapport 17; ICW: Wageningen, The Netherlands, 1963.
50. PTG. Polish Soil Classification, 5th ed.; Soil Science Annual; 2011; Volume 62, p. *193*. Available online: http://ssa.ptg.sggw.pl/en/artykul/2810/polish-soil-classification-fifth-edition (accessed on 13 December 2020). (In Polish).
51. FAO; IUSS Working Group WRB. *World Reference Base for Soil Resources 2014. International Soil Classification System for Naming Soils and Creating Legends for Soil Maps*; World Soil Resources Reports, No. 106; FAO: Rome, Italy, 2015; p. 203. ISBN 978-92-5-108369-7.
52. Schaap, M.G.; Leij, F.J.; van Genuchten, M.T. Neural network analysis for hierarchical prediction of soil water retention and saturated hydraulic conductivity. *Soil Sci. Soc. Am. J.* **1998**, *62*, 847–855. [CrossRef]
53. Schaap, M.G.; van Genuchten, M.T. A modified Mualem-van Genuchten formulation for improved description of the hydraulic conductivity near saturation. *Vadose Zone J.* **2006**, *5*, 27–34. [CrossRef]
54. Rumsey, D.J. Statistics for Dummies, 2nd ed.; 2016; 408p. Available online: https://www.gettextbooks.com/author/Deborah_J_Rumsey (accessed on 19 December 2020).
55. Zhang, X.; Zhu, J.; Wendroth, O.; Matocha, C.; Edwards, D. Effect of macroporosity on pedotransfer function estimates at the field scale. *Vadose Zone J.* **2019**, *18*, 180151. [CrossRef]
56. Nash, J.E.; Sutcliffe, J.V. River Flow Forecasting through Conceptual Model. Part 1—A Discussion of Principles. *J. Hydrol.* **1970**, *10*, 282–290. [CrossRef]
57. Mulla, D.J.; Mc Bratney, A.B. Soil spatial variability. In *Soil Physics Companion*; Warrick, A.W., Ed.; CRC Press: Boca Raton, FL, USA, 2002; pp. 343–373.
58. Borek, Ł. The use of different indicators to evaluate chernozems fluvisols physical quality in the Odra River valley: A case study. *Pol. J. Environ. Stud.* **2019**, *28*, 4109–4116. [CrossRef]
59. Rubio, C.M.; Poyatos, R. Applicability of HYDRUS-1D in a Mediterranean mountain area submitted to land use changes. *Int. Sch. Res. Not. Soil Sci.* **2012**, 1–7. Available online: http://downloads.hindawi.com/archive/2012/375842.pdf (accessed on 13 December 2020). [CrossRef]
60. Gill, S.M. Temporal Variability of Soil Hydraulic Properties under Different Soil Management Practices. 2012. Available online: https://www.researchgate.net/publication/311674921_ (accessed on 20 December 2020).
61. Pollacco, J.A.P.; Webb, T.; McNeill, S.; Hu, W.; Carrick, S.; Hewitt, A.; Lilburne, L. Saturated hydraulic conductivity model computed from bimodal water retention curves for a range of New Zealand soils. *Hydrol. Earth Syst. Sci.* **2017**, *21*, 2725–2737. [CrossRef]

62. Dexter, A.R.; Czyż, E.A.; Gaţe, O.P. Soil structure and the saturated hydraulic conductivity of subsoils. *Soil Tillage Res.* **2004**, *79*, 185–189. [CrossRef]
63. Rezaei, M.; Saey, T.; Seuntjens, P.; Joris, I.; Boënne, W.; Van Meirvenne, M.; Cornelis, W. Predicting saturated hydraulic conductivity in a sandy grassland using proximally sensed apparent electrical conductivity. *J. Appl. Geophys.* **2016**, *126*, 35–41. [CrossRef]
64. Merdun, H.; Çınar, Ö.; Meral, R.; Apan, M. Comparison of artificial neural network and regression pedotransfer functions for prediction of soil water retention and saturated hydraulic conductivity. *Soil Tillage Res.* **2006**, *90*, 108–116. [CrossRef]
65. García-Gutiérrez, C.; Pachepsky, Y.; Martín, M.A. Technical note: Saturated hydraulic conductivity and textural heterogeneity of soils. *Hydrol. Earth Syst. Sci.* **2018**, *22*, 3923–3932. [CrossRef]
66. Hillel, D. *Fundamentals of Soil Physics*; Academic Press: New York, NY, USA, 2013; 413p, ISBN 9780080918709.

 *land*

Article

Concept of Soil Moisture Ratio for Determining the Spatial Distribution of Soil Moisture Using Physiographic Parameters of a Basin and Artificial Neural Networks (ANNs)

Edyta Kruk * and Wioletta Fudała

Department of Land Reclamation and Environmental Development, Faculty of Environmental Engineering and Land Surveying, University of Agriculture in Krakow, Al. Mickiewicza 24-28, 30-059 Kraków, Poland; wioletta.zarnowiec@urk.edu.pl
* Correspondence: edyta.kruk@urk.edu.pl

Abstract: The results of investigations on shaping the soil moisture ratio in the mountain basin of the Mątny stream located in the Gorce region, Poland, are presented. A soil moisture ratio was defined as a ratio of soil moisture in a given point in a basin to the one located in a base point located on a watershed. Investigations were carried out, using a TDR device, for 379 measuring points located in an irregular network, in the 0–25 cm soil layer. Values of the soil moisture ratio fluctuated between 0.75 and 1.85. Based on measurements, an artificial neural network (ANN) model of the MLP type was constructed, with nine neurons in the input layer, four neurons in the hidden layer and one neuron in the output layer. Input parameters influencing the soil moisture ratio were chosen based on physiographic parameters: altitude, flow direction, height a.s.l., clay content, land use, exposition, slope shape, soil hydrologic group and place on a slope. The ANN model was generated in the module data mining in the program Statistica 12. Physiographic parameters were generated using a database, digital elevation model and the program ArcGIS. The value of the network learning parameter obtained, 0.722, was satisfactory. Comparison of experimental data with values obtained using the ANN model showed a good fit; the determination coefficient was 0.581. The ANN model showed a minimal tendency to overestimate values. Global network sensitivity analysis showed that the highest influence on the wetness coefficient were provided by the parameters place on slope, exposition, and land use, while the parameters with the lowest influence were slope, clay fraction and hydrological group. The chosen physiographic parameters explained the values of the relative wetness ratio a satisfactory degree.

Keywords: soil moisture; physiographic parameters of basins; artificial neural network (ANN); redundancy analysis (RDA)

Citation: Kruk, E.; Fudała, W. Concept of Soil Moisture Ratio for Determining the Spatial Distribution of Soil Moisture Using Physiographic Parameters of a Basin and Artificial Neural Networks (ANNs). *Land* **2021**, *10*, 766. https://doi.org/10.3390/land10070766

Academic Editor: Krish Jayachandran

Received: 22 May 2021
Accepted: 17 July 2021
Published: 20 July 2021

Publisher's Note: MDPI stays neutral with regard to jurisdictional claims in published maps and institutional affiliations.

Copyright: © 2021 by the authors. Licensee MDPI, Basel, Switzerland. This article is an open access article distributed under the terms and conditions of the Creative Commons Attribution (CC BY) license (https://creativecommons.org/licenses/by/4.0/).

1. Introduction

Soil moisture [1,2], especially in the context of climate change, is one of the greatest problems in hydrological responses [3,4], agriculture and on industrial sites [5–9]. The question of the influence of initial soil moisture, despite being mentioned by many studies as a significant erosion factor, has rather rarely been extensively researched [10]. The actual state of soil moisture determines the beginning of surface runoff. Surface runoff is shaped by many phenomena, including exceeding a soil's full saturation and filtration possibilities, subsurface outflow, and groundwater filtration [11,12]. The spatial distribution of soil moisture is realized by various methods [13], such as: direct measurements; the use of models based on physiographic basin parameters [14] as well as remote sensing data [15,16]; and in many cases, a combination of the above methods; and has been the subject of interest of many authors.

Gómez-Plaza et al. [17] carried out investigations of moisture distribution in a small basin located in Spain, using 50 sample points over a 20 m × 20 m regular network.

They used a TDR (time domain reflectometry) device to measure the soil moisture at depths of 15 cm. They connected the spatial distribution of soil moisture with use of TWI (topographic wetness index). They also pointed out a use for these indicators for rather wet climates. TWI is calculated by the following equation:

$$TWI = ln\left(\frac{\alpha}{tan\beta}\right) \tag{1}$$

where: α is the specific catchment area defined as the local upslope area draining through a unit contour length, which is equal to the grid cell width in this study; and β is the local slope gradient [18].

Tombul [19], simulating surface runoff depending on soil initial moisture, introduced shaping of the spatial distribution in the upper soil layer. The investigations were carried out in the Kurukavak river basin, located in Turkey, in a 4.25 km^2 area. Soil moisture measurements were carried out by means of a TDR device. For investigation of spatial distribution, he used the Xinanjiang model [20], based on basic hydrological responses such as: evapotranspiration, runoff generation and flow routing, connecting soil moisture distribution with wetness index (WI):

$$\theta_{WI} = \theta_s \cdot \frac{1 - \left(1 - \frac{\theta}{\theta_s}\right)^{\frac{1}{1+b}}}{1 - (1 - WI)^{\frac{1}{b}}} \tag{2}$$

where: θ_{WI} is the critical value of capacity, θ_s is the moisture at full saturation, $\theta_{WI} \leq \theta_s$, θ is the actual moisture, b is the shape parameter.

For his investigations, he used high-resolution DEM, constructed based on a topographic map, using the vector model TIN (triangulated irregular network). The introduced distribution was substituted by a topographic index.

Zhang et al. [21] researched the spatial distribution of soil moisture at the 10-cm layer, on a 9.8-ha experimental plot, using a 15 m × 15 m network comprising 250 points in total. They showed that moisture was characterized by the normal distribution. Spatial distribution was examined using kriging, using the module Spatial Analyst connected in the program ArcGIS release 9.0. They showed the significance of investigations of moisture distribution for fertilization and irrigation needs.

Merdun et al. [22] studied the spatial distribution of water retention, for three various initial moistures (dry, mean, and wet), obtained using a rainfall simulator on soil at the experimental centre of the University of Sutcu Imam in Turkey, using 100 sampled points. The structure of the spatial distribution and the semivariogram were executed in the program GS+ 5.1.

Penna et al. [23] investigated moisture distribution in the upper soil layer, in a 1.9 km^2 area located in a mountain basin in the Italian Alps, using 42 sample points. They showed the significance of relief parameters and atmospheric conditions, including slope, TWI, exposition, and solar radiation for the shaping of this property, showing a 42% share of these factors to explain moisture distribution.

Temimi et al. [24] studied the distribution of soil moisture in the Mackenzie River basin, over a 4000 km^2 area located in Canada. In their investigations, they used a modified, not-statistical topographic wetness index (TWI), determined based on DEM:

$$TWI = ln(a) - ln(\alpha \cdot tan(\beta)) \cdot e^{-u\,LAI}) \tag{3}$$

where: a is the alimentation area, β is the local slope, LAI is the leaf area index, μ is the coefficient of radiation extinction, depending on plant cover, with values between 0.35 and 0.70.

DEM was determined in the framework of the program Shuttle Radar Topography Mission (SRTM). LAI was in turn determined from satellite observations.

Fan et al. [25] carried out investigations on spatial variability of soil moisture, using a regular 50 m network with 20 sample points, on an experimental field in China. They used kriging, realized in the program ArcGIS release 9.0.

Nikolopulos et al. [11] carried out investigations on the influence of initial soil moisture in the Fella river basin, including an area of 623 km^2 and their subbasins, located in the Italian Alps, using peak flow and outflow discharge during a rainstorm. For investigation of the spatial distribution of soil moisture, they used the hydrologic model Triangulated Irregular Network (TIN), based on the Real-Time Integrated Basin Simulator. They used a DEM with a resolution of 20 m.

Jia et al. [26] examined the temporary stability of spatial distribution of moisture in loess soil in China, on five experimental plots of dimension 61 × 5 m, in 10 cm intervals, up to a depth of 1 m. They showed that the vertical distribution of moisture does not have a unitary trend; however, the horizontal distribution was similar for the same layers, irrespective of a density increase.

This work aims to present the concept of determination of soil moisture distribution, based on physiographic parameters of a basin, realized by the use of artificial neural networks (ANNs).

2. Materials and Methods

2.1. Study Sites

The study was carried out in the Outer Western Carpathians, in the southern region of Małopolska Province, Poland (Figure 1). The Mątny Stream (1.47 km^2) flows into the Mszanka River in the hamlet of Skiby at 20°9′2.35″ E, 49°37′30.52″ N. The main watercourse starts at 20°08′28.52″ E, 49°36′25.44″ N. The mean annual air temperature is 7.4 °C. The lowest temperature of −25.8 °C was recorded on 3 February 2012 and the highest, 33.0 °C, on 8 July 2013. The long-term annual average total precipitation is 846.63 mm and the total precipitation over the years in the study was 948.1 mm. The highest daily precipitation was 100.7 mm on 15 May 2014. The snow usually starts to melt at the beginning of March. The growing season starts around 10 April, and the winter begins around 30 November.

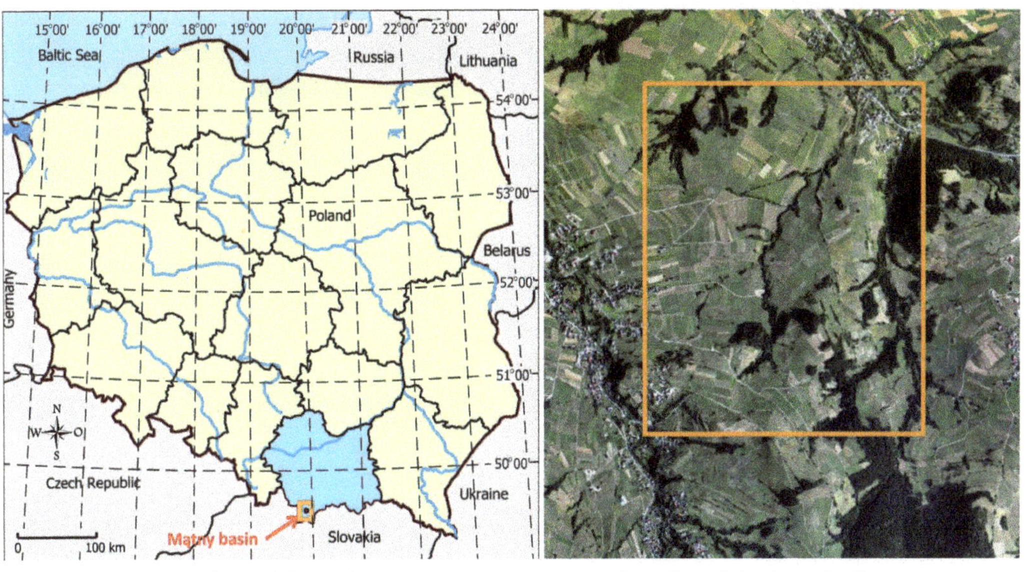

Map of Poland

Location of the investigation area

Figure 1. Location of the research area within Poland.

The terrain comprises low- and medium-height mountains with peak heights ranging from 617.6 m above sea level (m.a.s.l.) to 732.0 m.a.s.l. The lowest point is situated 490.0 m.a.s.l. The mean height of the catchment is 582.66 m.a.s.l. [27]. The slope distribution is as follows: <5%, (0.06 km^2); 5–10% (0.27 km^2); 10–18% (0.67 km^2); 18–27% (0.31 km^2); and >27% (0.16 km^2). The weighted average slope for the entire catchment was 16.28% [28]. The catchment land use structure was dominated by grassland (73.5%); arable lands constituted 14.3% and included the following crops: spring oats (*Avelana sativa*), 7.3%; potatoes (*Solanum tuberosum*), 4.3%; and common wheat (*Triticum aestivum*), 2.7%. Forests accounted for 9.5% and urban areas for 2.7% of the catchment land (Figure 2). The catchment area is cut by a network of dirt roads. Most of them are deeply furrowed and tend to transform into water-carrying streams during and after rain events.

Figure 2. Photos showing typical land use of the Mątny basin.

Soils in the catchment area are diverse, depending on slope location. The soil cover in the Mątny stream catchment is dominated by loamy soils, including sandy clay loam, loam, silt loam, clay loam and sandy loam. Pedological conditions were identified by analysis of a 1:25,000 agricultural soil map and categorized into the respective groups according to USDA standards [29].

2.2. Experimental Design/Models Applied

Calculations implementing the soil moisture ratio (wetness coefficient) model algorithm were performed using the interface of ArcGIS 10.3.1 software [30]. The spatial data were obtained based on the Polish State Surveying Coordinate System (PUGW). The following parameters were determined: altitude, slope, flow direction, exposition, shape of the slope, situation on the slope, hydrological group, use, and clay fraction. Thematic layers were developed using the sources specified in Table 1.

Table 1. Description of parameters used in coefficient wetness model.

Parameter	Explanation	
DEM	Geodesic and Cartographic Documentation Centre	Scale 1:5000, resolution 5 m
Altitude	meters mean sea level	m.a.s.l.
Slope	Differences of heights between the points Δh divided by length of projection of direction between the points, l; [%] [1]	Intervals, %:0–5; 5–10; 10–18; 18–27; >27
Flow direction	Determined based on height difference between the given cell determined each of the eight adjacent cells, based on one-direction points model D8 [2] [31], where: Z is the number of adjacent cells, h is the resolution of the GRID model, hØ(i) is the distance between the middle points of cell, 1 for the ones situated in the cardinal directions (N,E,S,W), root square for the two remaining ones.	Convention of the DEM numbering (a) in standard cell system, (b) in determination of flow directions, (c) coding of flow directions by the D8 algorithm [31]
Exposition	Location on the slope with respect to the direction of sunlight rays, determined using a 4-grade scale, as one of the geographical directions.	East–E, west–W, north–N, south–S
Shape of the slope	Concave, flat, convex	
Situation on a slope	Determined using a five-grade scale: hilltop, slope outset, slope middle, slope bottom, slope foot	Place on slope: 1- hilltop 2- slope outset 3- slope middle 4- slope bottom 5- slope foot
Digital soil map	Institute of Soil Science and Plant Cultivation State Research Institute	Scale 1:25,000
Hydrological soil group	Determined based on U.S. Department of Agriculture–Natural Resources Conservation Service [32] method	Groups:A, B, C, D
Use	Orthophoto map; site inspection	Scale 1:1000Data: 25.07.2014
Clay fraction	Determined using the Casagrande'a method. Soil samples were collected from a top layer of the soil (0 to 10 cm) in 43 selected study plots.	

where: [1] is $S = \frac{\Delta h}{l} \cdot 100$; [2] is $S_{D8} = max_{i=1,8} \frac{Z_9 - Z_i}{h\emptyset(i)}$.

Three sampling time periods were adopted for each of the three growing seasons: dry (period I), medium (period II) and wet (period III) AMC (Antedescent Moisture Conditions), distinguished based on the sum of rainfall from the previous 5 days (mm) and the season category (dormant or growing) (Table 2). In botany and agriculture, the growing season is defined as the portion of each year when native plants and ornamental plants grow, while the dormant season is when growth and development are temporarily stopped [33]. In much of Europe, the growing season is defined as the average number of days with a 24-h average temperature of at least 6 °C; in southern Poland, this typically lasts from April to September.

Table 2. AMC classes [34].

AMC Classes	Vegetative Dormant Season	Vegetative Growing Season
I	Less than 12.7	Less than 35.5
II	12.7 to 28.0	35.5 to 53.3
III	More than 28	More than 53.3

The soil moisture ratio $K_{w,i}$ in point i was determined based on the concept of wetness coefficient [35,36]:

$$K_{w,i} = \frac{\theta_i}{\theta_b} \tag{4}$$

where: θ_i is the soil volumetric water in the given point i within the basin area [m^3·m^{-3}], and θ_b is the soil volumetric water content in the basal point [m^3·m^{-3}].

From this, the soil moisture in a point i in a basin can by determined as:

$$\theta_i = \theta_b \cdot K_{w,i}(p_1,\ p_2 \ldots p_n) \tag{5}$$

where: $K_{w,i}(p_1,\ p_2 \ldots p_n)$ is the relative wetness coefficient determined based on the following parameters: place on a slope, exposition, land use, shape of slope, altitude, flow direction, slope, clay fraction, hydrological group.

The purpose of the wetness coefficient was to assess the distribution of soil moisture in the catchment area, based on the measurement from a basal point located on a hilltop. Soil moisture was measured in 2014, between 11:00 pm and 14:00 am during days without and with rainfall using the TDR device, as a mean value in the 0–0.10 m layer, for 379 measuring points (153 for period I, 100 for period II and 126 for period III). The distribution of points was random, taking into account the specific character of the mountain catchment. The measurements were made, for period I, on 25 July (the sum of rainfall from the previous 5 days was 21.2 mm, in the vegetative growing season); for period II, on 4 October (the sum of rainfall from the previous 5 days was 21.7 mm, in the growing season); and for period III, on 25 September (the sum of rainfall from the previous 5 days was 34.8 mm, in the dormant season) (Figure 3).

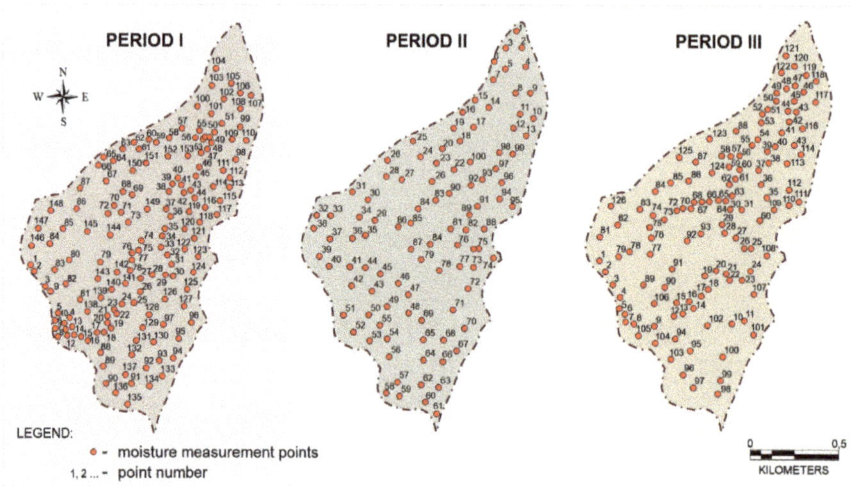

Figure 3. Location of moisture measurement points.

2.3. Statistical Analysis and Data Procedure

An ANN (artificial neural network) model in the form of a multilayer perceptron (MLP) was used to generate wetness coefficients, using Statistica software (release 12.5). A total of 70% of all variables were applied for the learning process; 15% were used for validation and 15% to test the model. A quasi-Newton algorithm with a Broyden–Fletcher–Goldfarb–Shanno (BFGS) modification was selected for the learning neural network. The sum of squares (SOS) was treated as the error function. The artificial neural network model was used to establish the association between the wetness coefficient, whereas a physiographic parameter, indicative of the soil and its use, was applied in all data set as an independent variable. The multi-layer perceptron consisted of three layers of neurons: (1) an input layer, (2) an output layer and (3) intermediate (hidden) layers. Each neuron had a number of inputs (from outside to the subsequent layer or out of the network) [37]. In this study, the network system included an input and a hidden layer made of nine neurons (altitude, slope, flow direction, exposition, shape of the slope, situation on a slope, hydrological group, land use and clay fraction) and an output layer with one neuron (wetness coefficient).

Studies on the percentage effect of selected soil, use and physiographic parameters were carried out using ANN based on a global network sensitivity analysis [38–40].

RDA (redundancy analysis) of standardized environmental variables was performed to explain and describe the pattern of variability in a parameter [41] Multivariate analysis showed the presence of two main gradients of environmental variables. Positions of the vectors of independent variables were proportional to the loading factors. The multivariate analysis was carried out with Canoco for Windows version 4.51.

The analysis of empirical model adjustment to experimental data was carried out using the following metrics [42], presented in Table 3.

Table 3. Measures of model performance [43,44].

Measure	Equation
Mean error of prediction	MEP [1]
Root mean square error	RMSE [2]
Mean percentage error	MPE [3]
Model efficiency	ME [4]

where: [1] is $MEP = \frac{1}{n} \cdot \sum_{i=1}^{n} \left(c_i^m - c_i^p \right)$; [2] is $RMSE = \sqrt{\frac{1}{n} \cdot \sum_{i=1}^{n} \left(c_i^m - c_i^p \right)^2}$; [3] is $MPE = \frac{1}{n} \cdot \sum_{i=1}^{n} \frac{c_i^m - c_i^p}{c_i^m} \cdot 100\%$;

[4] is $ME = 1 - \frac{\sum_{i=n}^{n} \left(c_i^m - c_i^p \right)^2}{\sum_{i-n}^{n} \left(c_i^m - \bar{c} \right)^2}$; c_i^m is the measured values, c_i^p is the computed values, and n is the number of data.

Spatial variability of the investigated and measured values of the soil moisture ratio was determined using kriging. The kriging method allows to obtain the most probable values in any part of an investigated area and to find the location for new measuring points. To briefly define the technique of kriging, it is a method for optimizing the estimation of the spatially correlated quantity Z, both in stationarity and non-stationarity instances.

3. Results

Moisture at the base point was $0.146\ \mathrm{m^3 \cdot m^{-3}}$ in the dry season (period I), $0.247\ \mathrm{m^3 \cdot m^{-3}}$ in the medium season (period II), and $0.433\ \mathrm{m^3 \cdot m^{-3}}$ in the wet season (period III). The measured moisture fluctuated between $0.086\ \mathrm{m^3 \cdot m^{-3}}$ and $0.226\ \mathrm{m^3 \cdot m^{-3}}$ for period I, between $0.146\ \mathrm{m^3 \cdot m^{-3}}$ and $0.354\ \mathrm{m^3 \cdot m^{-3}}$ for period II, and between $0.145\ \mathrm{m^3 \cdot m^{-3}}$ and $0.441\ \mathrm{m^3 \cdot m^{-3}}$ for period III. The obtained values were in accordance with the ones introduced in the literature [45] The ANN model was then generated based on the results of the investigation of physiographic parameters (Figure 4).

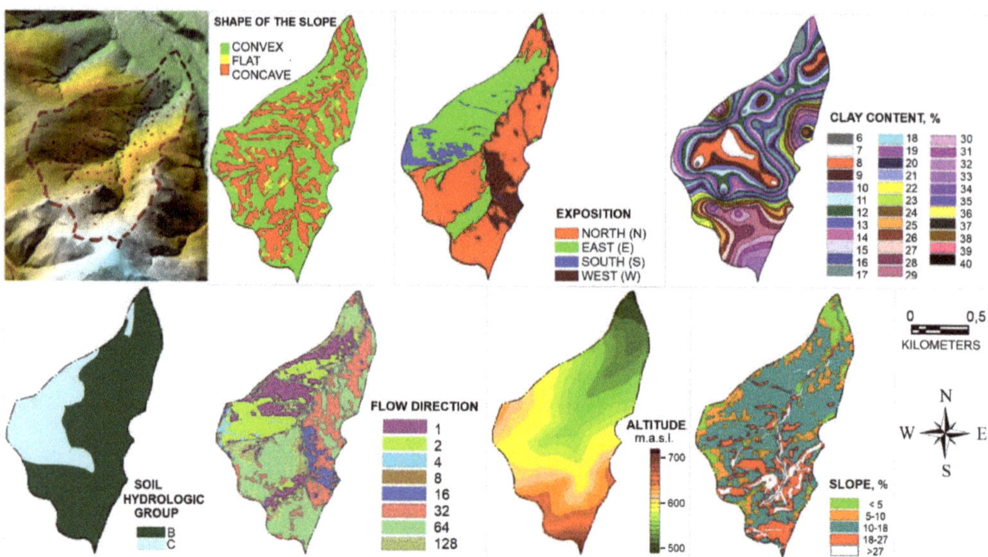

Figure 4. Physiographical parameters of the Mątny stream basin.

The best-fitting model turned out to be the MLP 9-4-1, with four perceptrons in the hidden layer. The MLP 9-4-1 model is characterized by high fitting quality and low error, and therefore presents a good adjustment (Table 4). Thanks to the global sensitivity analysis (Table 5), the relative and absolute influence of a number of parameters on the soil moisture ratio K_w were determined.

Table 4. Analysis of quality and errors of the MLP 9-4-1.

Quality	learning	0.757
	testing	0.752
	validation	0.807
Error	learning	0.006
	testing	0.009
	validation	0.004
Perceptron activation functions	hidden	
	output	

Table 5. Global sensitivity analysis of the MLP 9-4-1 model.

No.	Parameter	Relative Sensitivity	Absolute Influence [%]
1	Place on slope	11.3	40
2	Exposition	3.5	12
3	Use	3.0	11
4	Shape of the slope	2.9	10
5	Altitude	2.9	10
6	Flow direction	1.6	5.0
7	Slope	1.3	5.0
8	Clay fraction	1.0	4
9	Hydrological group	1,0	4

In the Mątny stream basin, the parameters place on the slope (40%) and exposition (12%) had the highest impact (Table 5). Values of the soil moisture ratio fluctuated between 0.75 and 1.85. Based on the simulated values of the soil moisture ratio for every one of the measured points, the map of the spatial distribution of this parameter was generated using Surfer 10 and ArcGIS, based on data generated by ANN 9-4-1 (Figure 5a) and on the measurement date (Figure 5b). The highest values of the soil moisture ratio occurred in the northwest part of the basin. Figure 6 presents a comparison between the ANN simulated values of the soil moisture ratio and the one elaborated based on measured data. Values of the simulated wetness coefficient ranged between 0.89 and 1.13. Model efficiency measures for the MLP 9-4-1 were as follows: MEP: 0.004, RMSE: 0.104, MPE: −0.6%, ME: 0.580 and R^2: 0.581 (Table 6).

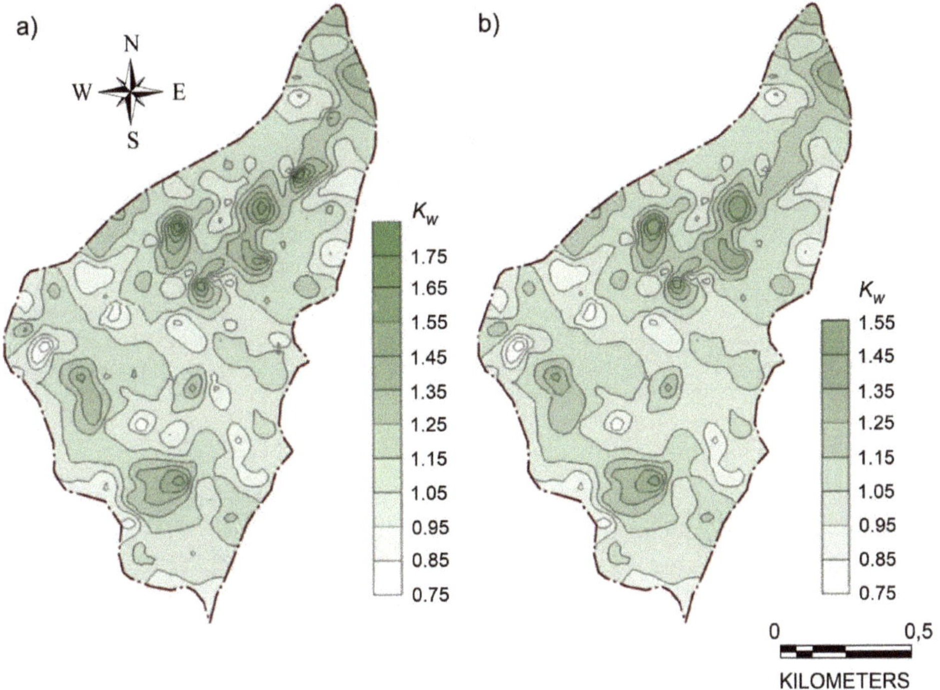

Figure 5. Spatial distribution of the soil moisture ratio in the Mątny basin generated using: (**a**) the MLP 9-4-1 model, and (**b**) based on measured data elaborated by kriging technique.

Table 6. Model efficiency measures for the MLP 9-4-1.

Model Efficiency Measures.				
MEP	RMSE	MPE [%]	ME [-]	R^2 [-]
0.004	0.104	−0.6	0.580	0.581

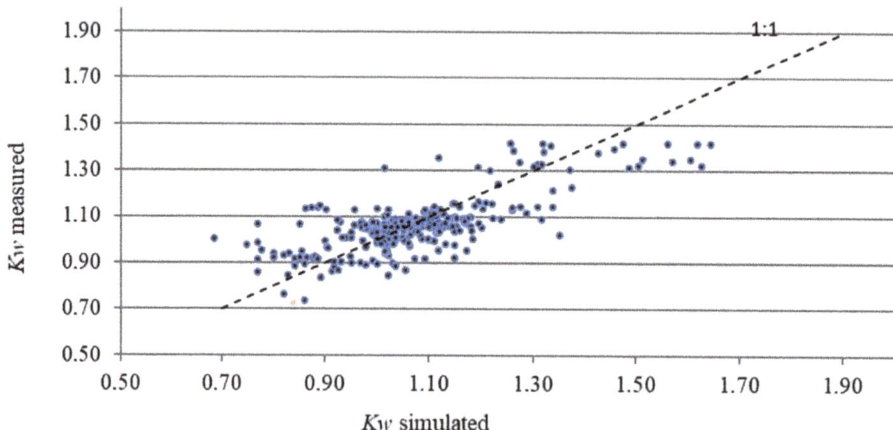

Figure 6. The measured versus simulated soil moisture ratio.

Multivariate analysis (RDA) was employed to explain the relative importance of particular explanatory variables and underline differences between parameters. The patterns of the independent variables for soil factors and environmental parameters are shown on the plot (Figure 7). We noted a positive correlation among the flow direction, slope, clay content and wetness coefficient. Generally, multivariate analysis for the studied parameters indicated a strong positive correlation among the soil hydrologic group. Moreover, we saw a correlation between altitude and place on the slope. On the other hand, only exposition N, exposition E and exposition W were strongly negatively correlated to other examined parameters. The first component described 27.43% of the total variation, and the second component explained 42.01% of the total variance among the study parameters.

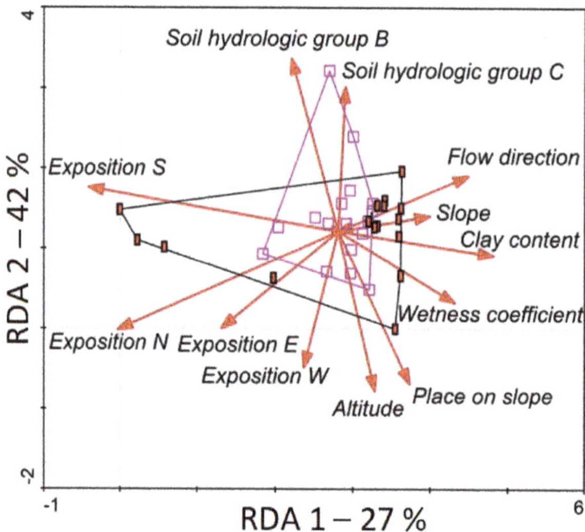

Figure 7. Multivariate RDA analysis for the soil moisture ratio is created by vectors (red squares are for convex; purple squares for concave slopes).

4. Discussion

In this study, the concept of relative wetness coefficient was developed as a base for determination of the spatial distribution of soil moisture on an area of a mountain basin.

The idea of the coefficient is based on the observations that there is a relation between soil moisture in a point on a slope and on a flat in a basin. Values of relative wetness coefficient were obtained by the ANN model generated based on nine basin parameters: place on a slope, exposition, land use, shape of slope, altitude, flow direction, slope, clay fraction and hydrological group and direct measurements of soil moisture on the area of small Carpathian basin (1.47 km^2). Parameters of ANN learning, testing, validation, and model efficiency measures were very satisfactory. The model overestimated prognosed values only of 0.6%. Model was fitted to the experimental data very well. Based on the ANN model, the sequence of significance of the particular basin parameters were arranged to provide the relative wetness coefficient explanation. The highest influence had: place on slope, exposition, and land use. In turn, clay fraction and the hydrological group had no significance, because the investigated basin is not highly differentiated regarding soil. A very interesting model for evaluation of the relative wetness coefficient foe small basin was presented by Svetlitchnyi et al. [33]. It was based on basin parameters such as: the shape of slope, distance from the divide, slope aspect and overall length of a slope. The obtained model errors were very satisfactory. The relative wetness coefficient can be used as an easy way to determine the distribution of soil moisture in the upper layer in small mountain basins.

5. Conclusions

The soil moisture ratio, as a relation between soil moisture in a given point and in a basal point located on a flat, generated based on basin parameters, by use of artificial neural networks is one of the simplest ways to determine the distribution of soil moisture on an area of small mountain basin. The main basin parameters that can influence the relative soil moisture ratio are the place on a slope, exposition, land use (if it is not uniform on an area of a basin), shape of slope and altitude. Our investigations showed that they controlled the relative soil moisture ratio in sum in 83%. The model for determination of the relative soil moisture ratio, generated by use of the ANN, gave very satisfied results in terms of errors, and explained as many as 58% of cases. The ANN model overestimated prognosis values only of 0.6%. For larger basins and basins with highly differentiated soil, the next studies have to be carried out.

Author Contributions: Conceptualization, E.K. and W.F.; methodology, E.K.; software, E.K.; validation, E.K. and W.F.; formal analysis, E.K.; investigation, E.K. and W.F.; resources, E.K. and W.F.; data curation, E.K. and W.F.; writing—original draft preparation, E.K.; writing—review and editing, E.K. and W.F.; visualization, E.K. and W.F.; supervision, E.K.; project administration, E.K.; funding acquisition, E.K. and W.F. Both authors have read and agreed to the published version of the manuscript.

Funding: This research was funded by the Department of Land Reclamation and Environmental Development, Faculty of Environmental Engineering and Land Surveying, University of Agriculture in Krakow.

Institutional Review Board Statement: Not applicable.

Informed Consent Statement: Not applicable.

Data Availability Statement: Not applicable.

Conflicts of Interest: The authors declare no conflict of interest.

References

1. Li, J.; Islam, S. Estimation of root zone soil moisture and surface fluxes partitioning using near surface soil moisture measurement. *J. Hydrol.* **2002**, *259*, 1–14. [CrossRef]
2. Romana, N. Soil moisture at local scale: Measurements and simulations. *J. Hydrol.* **2014**, *516*, 6–20. [CrossRef]
3. Castillo, V.M.; Gómez-Plaza, A.; Martinez-Mena, M. The role of antecendent soil water content in the runoff response of a semiarid catchment: A simulation approach. *J. Hydrol.* **2003**, *284*, 114–130. [CrossRef]
4. Western, A.W.; Zhou, S.L.; Grayson, R.B.; McMahon, T.A.; Blöschl, G.; Wilson, D.J. Spatial correlation of soil moisture in small catchments and its relationship to dominant spatial hydrological processes. *J. Hydrol.* **2004**, *286*, 113–134. [CrossRef]

5. Boroń, K.; Klatka, S.; Ryczek, M.; Liszka, P. Kształtowanie się właściwości fizycznych, fizykochemicznych i wodnych rekultywowanego i niezrekultywowanego osadnika byłych Krakowskich Zakładów Sodowych "Solvay"/The formation of the physical, physico-chemical and water properties reclaimed and not reclaimed sediment reservoir of the former Cracow Soda Plant "Solvay". *Acta Sci. Pol. Form. Circumiectus* **2016**, *15*, 35–43. [CrossRef]

6. Brocca, L.; Morbidelli, R.; Melone, F.; Moramaraco, T. Soil moisture spatial variability in experimental areas of central Italy. *J. Hydrol.* **2007**, *333*, 356–373. [CrossRef]

7. Klatka, S.; Malec, M.; Ryczek, M.; Boroń, K. Wpływ działalności eksploatacyjnej Kopalni Węgla Kamiennego "Ruch Borynia" na gospodarkę wodną wybranych gleb obszaru górniczego/Influence of mine activity of the coal mine "Ruch Borynia" on water management of chosen soils on mining area. *Acta Sci. Pol. Form. Circumiectus* **2015**, *14*, 115–125. Available online: http://www.formatiocircumiectus.actapol.net/tom14/zeszyt1/14_1_115.pdf (accessed on 24 February 2021). [CrossRef]

8. Klatka, S.; Malec, M.; Ryczek, M.; Kruk, E.; Zając, E. Ocena zdolności retencyjnych wybranych odpadów przemysłowych/Evaluation of retention ability of chosen industrial wastes. *Acta Sci. Pol. Form. Circumiectus* **2016**, *15*, 53–60. Available online: http://www.formatiocircumiectus.actapol.net/tom15/zeszyt4/15_4_53.pdf (accessed on 15 January 2021).

9. Wei, J.B.; Xiao, D.N.; Zeng, H.; Fu, Y.K. Spatial variability of soil properties in relation to land use and topography in a typical small watershed of the black soil region, northeastern China. *Environ. Geol.* **2008**, *53*, 1663–1672. [CrossRef]

10. Boron, K.; Klatka, S.; Ryczek, M.; Zając, E. Reclamation and cultivation of the Cracow soda plant lagoons, International Conference on Construction for a Sustainable Environment, Vilnius, LITHUANIA, JUL 01–04, 2008. In *Construction for a Sustainable Environment*, 1st ed.; Sarsby, R., Ed.; Routledge: London, UK, 2010; pp. 245–250. Available online: https://www.routledge.com/Construction-for-a-Sustainable-Environment/Sarsby-Meggyes/p/book/9780415566179 (accessed on 20 July 2021).

11. Nikolopulos, E.I.; Anagnostou, E.N.; Borga, M.; Vivoni, E.R.; Papadopoulos, A. Sensitivity of a mountain basin flash flood to initial wetness condition and rainfall variability. *J. Hydrol.* **2011**, *402*, 165–178. [CrossRef]

12. Santra, P.; Sankar, B.; Chakravarty, D. Spatial prediction of soil properties in a watershed scale through maximum likelihood approach. *Environ. Earth Sci.* **2012**, *65*, 2051–2061. [CrossRef]

13. Alvarez-Garreton, C.; Ryu, D.; Western, A.W.; Crow, W.T.; Robertson, D.E. The impacts of assimilating satellite soil moisture into a rainfall-runoff model in a semi-arid catchment. *J. Hydrol.* **2014**, *519*, 2763–2773. [CrossRef]

14. Kruk, E.; Klapa, P.; Ryczek, M.; Ostrowski, K. Influence of DEM Elaboration Methods on the USLE Model Topographical Factor Parameter on Steep Slopes. *Remote Sens.* **2020**, *12*, 3540. [CrossRef]

15. Tischler, M.; Garcia, M.; Peters-Lidard, C.; Moran, M.S.; Miller, S.; Kumar, S.; Geiger, J. A GIS framework for surface-layer soil moisture estimation combining satellite radar measurements and land surface modeling with soil physical property estimation. *Environ. Model. Softw.* **2007**, *22*, 891–898. [CrossRef]

16. Vidhya Lakshmi, S.; Jijo, J.; Soundariya, S.; Vishalini, T.; Kasinatha Pandian, P. A comparison of Soil Texture Distribution and Soil Moisture Mapping of Chennai Coast using Landsat ETM+ and IKONOS Data. CONFERENCE ON WATER RESOURCES, COASTAL AND OCEAN ENGINEERING (ICWRCOE 2015). *Aquat. Procedia* **2015**, *4*, 1452–1460. [CrossRef]

17. Gómez-Plaza, A.; Martinez-Mena, M.; Albaladejo, J.; Castillo, V.M. Factors regulating spatial distribution of soil water content in small semiarid catchments. *J. Hydrol.* **2001**, *253*, 211–226. [CrossRef]

18. Zhao, L.; Liu, Y.; Luo, Y. Assessing Hydrological Connectivity Mitigated by Reservoirs, Vegetation Cover, and Climate in Yan River Watershed on the Loess Plateau, China: The Network Approach. *Water* **2020**, *12*, 1742. [CrossRef]

19. Tombul, M. Mapping Field Surface Soil Moisture for Hydrological Modeling. *Water Resour. Manag.* **2007**, *21*, 1865–1880. [CrossRef]

20. Gou, W.; Wang, C.; Ma, T.; Zeng, X.; Yang, H. A distributed Grid-Xinanjiang model with integration of subgrid variability of soil storage capacity. *Water Sci. Eng.* **2016**, *9*, 97–105.

21. Zhang, C.; Liu, S.; Fang, J.; Tan, K. Research on the spatial variability of soil moisture based on GIS. In *Computer and Computing Technologies in Agriculture, Volume I, The International Federation for Information Processing, Vol. 258, Proceedings of the First IFIP TC 12 International Conference on Computer and Computing Technologies in Agriculture (CCTA 2007), Wuyishan, China, 18–20 August 2007*; Li, D., Ed.; Springer: Boston, MA, USA, 2008; pp. 719–727. [CrossRef]

22. Merdun, H.; Meral, R.; Riza Demirkiran, A. Effect of the initial soil moisture content on the spatial distribution of the water retention. *Eurasian Soil Sci.* **2008**, *41*, 1098–1106. [CrossRef]

23. Penna, D.; Borga, M.; Norbiato, D.; Dalla Fontana, G. Hillslope scale soil moisture variability in a steep alpine terrain. *J. Hydrol.* **2009**, *364*, 311–327. [CrossRef]

24. Temimi, M.; Leconte, R.; Chaouch, N.; Sukumal, P.; Khanbilvardi, R.; Brissette, F. A combination of remote sensing data and topographic attributes for the spatial and temporal monitoring of soil wetness. *J. Hydrol.* **2010**, *388*, 28–40. [CrossRef]

25. Fan, Y.; Zhang, C.; Fang, J.; Tian, L. Research on Regional Spatial Variability of Soil Moisture Based on GIS. In *Computer and Computing Technologies in Agriculture III, IFIP Advances in Information and Communication Technology, Proceedings of the Third IFIP TC 12 International Conference (CCTA 2009), Beijing, China, 14–17 October 2009*; Li, D., Zhao, C., Eds.; Springer: Berlin/Heidelberg, Germany, 2010; Volume 317, pp. 466–470. [CrossRef]

26. Jia, Y.H.; Shao, M.A.; Jia, X.X. Spatial pattern of soil moisture and its temporal stability within profiles on a loessial slope in northwestern China. *J. Hydrol.* **2013**, *495*, 150–161. [CrossRef]

27. Halecki, W.; Kruk, E.; Ryczek, M. Loss of topsoil and soil erosion by water in agricultural areas: A multi-criteria approach for various land use scenarios in the Western Carpathians using a SWAT model. *Land Use Policy* **2018**, *73*, 363–372. [CrossRef]

28. Halecki, W.; Kruk, E.; Ryczek, M. Evaluation of Soil Erosion in the Mątny Stream Catchment in the West Carpathians Using the G2 model. *Catena* **2018**, *164*, 116–124. [CrossRef]
29. Soil Survey Staff. *Soil Taxonomy*; USDA: Washington, DC, USA, 1975; pp. 17–754.
30. Hillier, A. Manual for Working with ArcGIS10. Unpublished Paper. Available online: http://works.bepress.com/amy_hillier/24/ (accessed on 13 December 2020).
31. Wilson, D.J.; Gallant, J.C. Digital terrain analysis. In *Terrain Analysis: Principles and Applications*; Wilson, D.J., Gallant, J.C., Eds.; John Willey & Sons, Inc.: New York, NY, USA, 2000; pp. 1–27.
32. USDA—Natural Resources Conservation Service. *National Soil Survey Handbook, Title 430-VI*; US Department of Agriculture, Natural Resources Conservation Service: Washington, DC, USA, 2002.
33. Silveira, M.C.F.; Oliveira, E.M.M.; Carvajal, E.; Bon, E.P.S. Nitrogen regulation of Saccharomyces cerevisiae invertase. Role of the URE2 gene. *Appl. Biochem. Biotechnol.* **2000**, *84–86*, 247–254. [CrossRef]
34. United States Department of Agriculture. Soil Conservation Service 1993. In *National Engineering Handbook*; Section 4, Hydrolohy; Washington, DC, USA. Available online: https://directives.sc.egov.usda.gov/OpenNonWebContent.aspx?content=43924.wba (accessed on 20 July 2021).
35. Svetlitchnyi, A.A.; Plotnitskiy, S.V.; Stepovaya, O.Y. Spatial distribution of soil moisture content within catchments and its modelling on the basis of topographic data. *J. Hydrol.* **2003**, *277*, 50–60. [CrossRef]
36. Kruk, E.; Ryczek, M.; Malec, M.; Klatka, S. Kształtowanie się współczynnika wilgotności gleby w zlewni potoku Mątny w Gorcach/Shaping of the wetness coefficient in the Mątny stream basin in the Gorce mountains. *Inżynieria Ekologiczna* **2016**, *50*, 99–105. [CrossRef]
37. Dawson, J.F.; Richter, A.W. Probing three-way interactions in moderated multiple regression: Development and application of a slope difference test. *J. Appl. Psychol.* **2006**, *91*, 917–926. [CrossRef]
38. Rajurkar, M.P.; Kothyari, U.C.; Chaube, U.C. Artificial neural networks for daily rainfall-runoff modelling. *Hydrol. Sci. J.* **2002**, *47*, 865–877. [CrossRef]
39. Caamaño, D.; Googwin, P.; Manic, M. Derivation of a bedload sediment transport formula using artificial neural networks. In Proceedings of the 7th International Conference on Hydroinformatics HIC, Nice, France, 4–8 September 2006; pp. 1–8.
40. Kentel, E. Estimation of river flow by artificial neural networks and identification of input vectors susceptible to producing unreliable flow estimates. *J. Hydrol.* **2009**, *375*, 481–488. [CrossRef]
41. Lepš, J.; Smilauer, P. *Multivariate Analysis of Ecological Data Using CANOCO*; Cambridge University Press: Cambridge, UK, 2003. [CrossRef]
42. Rahnama, M.B.; Barani, G.A. Application of rainfall-runoff models to Zard River catchments. *Am. J. Environ. Sci.* **2005**, *1*, 86–89. [CrossRef]
43. Nash, J.E.; Sutcliffe, J.V. River flow forecasting through conceptual models. Part I—A discussion of principles. *J. Hydrol.* **1970**, *10*, 282–290. [CrossRef]
44. Tiwari, A.K.; Risse, L.M.; Nearing, M.A. Evaluation of WEPP and its comparison with USLE and RUSLE. *Trans. ASAE* **2000**, *43*, 1129–1135. [CrossRef]
45. Gebregiorgis, M.F.; Savage, M.J. Field, laboratory and estimated soil-water content limits. *Water SA* **2006**, *32*, 155–162. [CrossRef]

Article

Rate of Fen-Peat Soil Subsidence Near Drainage Ditches (Central Poland)

Ryszard Oleszczuk [1], Ewelina Zając [2,*], Janusz Urbański [3] and Jan Jadczyszyn [4]

[1] Institute of Environmental Engineering, Warsaw University of Life Sciences—SGGW, Nowoursynowska 166 St., 02-787 Warsaw, Poland; ryszard_oleszczuk@sggw.edu.pl

[2] Department of Land Reclamation and Environmental Development, Faculty of Environmental Engineering and Land Surveying, University of Agriculture in Krakow, Al. Mickiewicza 21, 31-120 Krakow, Poland

[3] Institute of Civil Engineering, Warsaw University of Life Sciences—SGGW, Nowoursynowska 166 St., 02-787 Warsaw, Poland; janusz_urbanski@sggw.edu.pl

[4] Institute of Soil Science and Plant Cultivation—State Research Institute, Czartoryskich 8 St., 24-100 Puławy, Poland; janj@iung.pulawy.pl

* Correspondence: ewelina.zajac@urk.edu.pl

Abstract: This study analyzed design depths (t_o), post-subsidence depths (t), shallowing magnitudes ($d = t_o - t$) and ratio values (d/t) of 12 drainage ditches in a fragment of the drained Solec fen-peat (central Poland) over a period of 47 years between 1967 and 2014. A significant decrease of the designed depth of the ditches t_o was shown, from the average designed value of 0.97 m to their average depth after subsidence, $t = 0.71$ m. The ratio (d/t) of 0.41, which is associated with the degree of organic matter decomposition, indicated medium degree of peat decomposition. The average values of bank and bottom subsidence of the ditches during the analyzed period, 1967–2014, were 0.43 m and 0.17 m, respectively. The values of the average annual rate of land surface subsidence in the vicinity of the ditches were varied and within the range of 0.09 cm year^{-1} to 1.70 cm year^{-1}, with an average of 0.92 cm year^{-1}. Two linear empirical equations were proposed to calculate the amount of subsidence and the average annual rate of subsidence of peat soil surface near the drainage ditch route, based on the knowledge of the initial thickness of the peat deposit. The results of calculations using the equations proposed by the authors were compared with calculations of the same parameters using 10 equations published in the literature. The results obtained using the proposed equations were mostly larger than those calculated with literature-published equations.

Keywords: drained peat soils; ditch subsidence; empirical equations of peat subsidence; drainage ditches

Citation: Oleszczuk, R.; Zając, E.; Urbański, J.; Jadczyszyn, J. Rate of Fen-Peat Soil Subsidence Near Drainage Ditches (Central Poland). *Land* **2021**, *10*, 1287. https://doi.org/10.3390/land10121287

Academic Editors: Wiktor Halecki, Dawid Bedla, Marek Ryczek and Artur Radecki-Pawlik

Received: 11 October 2021
Accepted: 18 November 2021
Published: 24 November 2021

Publisher's Note: MDPI stays neutral with regard to jurisdictional claims in published maps and institutional affiliations.

Copyright: © 2021 by the authors. Licensee MDPI, Basel, Switzerland. This article is an open access article distributed under the terms and conditions of the Creative Commons Attribution (CC BY) license (https://creativecommons.org/licenses/by/4.0/).

1. Introduction

In excessively humid areas (e.g., river valleys), the high groundwater table and low air content in the soil are conducive to the formation of peat soils, which are characterized by a high organic carbon content. Peat deposits covering only approximately 4% of the world are estimated to accumulate approximately 30% of the global organic carbon of all soils in their deposits [1–3]. In Europe, about 16% of peat soils are used for agricultural purposes (arable land and grassland), including the vast majority in Western European countries [4]. The area of organic soils and soils of organic origin in Poland estimated on the basis of the soil-agricultural map at a scale of 1:25,000 occupies 13.3% of agricultural land, wherein 5.6% are peat soils, 1.6% are sited peat soils, and 6.1% are moorsh soils [5]. These areas, used by many disciplines (including agriculture, horticulture, and forestry), have been drained in numerous cases [6]. This has resulted in the interruption of the existing organic matter accumulation process due to incomplete decomposition under conditions of very high humidity and lack of air access. In their natural state, these soils absorb carbon dioxide as a result of the assimilation and photosynthesis processes of the vegetation living in these areas. The change in air–water relations that accompanies the drainage of peatlands has caused the initiation of mucking and mineralization processes in organic matter, and the conversion of organic forms of nitrogen into mineral compounds and emission of greenhouse gases, such as carbon dioxide and nitrous oxide, into the atmosphere [7–9]. The ongoing climate change, significant increase in air temperature, and reduction in precipitation during the growing season lead to extreme weather phenomena, as well as increased frequency of agricultural drought [10–12]. The foregoing weather phenomena contribute to the intensification of mucking processes and permanent degradation of drained peat soils. Changes occur in their physical properties, i.e., retention, hydraulic properties, and soil compaction due to the shrinkage process, as do changes in their chemical properties, specifically reduction in organic carbon content and increase in ash content. All these phenomena contribute to the lowering of the surface (shallowing of the profile) in drained peat deposits, i.e., subsidence. This leads to a decrease in the thickness, as well as surface area, of these deposits and a permanent reduction in their agricultural production potential. It may even contribute to the total disappearance of peat soils in the natural environment and their transformation into typical mineral soils in the long run. This is an extremely unfavorable phenomenon, taking into consideration the numerous functions that peatlands play in the natural environment (e.g., accumulation of organic carbon, water retention, and biodiversity).

In the past, using peatlands for agricultural purposes (mainly meadows and pastures) required lowering the groundwater level to reduce moisture in the root zone (0–30 cm) in view of the water needs of grassland vegetation. These drainages were done in stages by constructing a network of deep draining channels, usually followed by a dense network of drainage ditches after several years (typically spaced about 100 m apart) and, as needed, an additional network of drainage pipes (spaced about 30 m apart). The foregoing drainage network parameters depend on the drainage depth (usually about 0.8–1.0 m) and the hydraulic properties of the peat soils, specifically the filtration coefficient [13]. Drainage of these soils for agricultural purposes, as opposed to drainage of mineral soils, has resulted in many negative physical, chemical, and biological processes, due to a decrease in the moisture content of the top layers, and a consequent increase in air content. Among the physical processes, the shrinkage process is observed due to a decrease in the moisture content of these soils, which consequently leads to increased compaction, reduced retention capacity, and changes in the hydraulic properties of these soils [14–19]. The disappearance of the buoyancy force and the pressure of the drained peat layers on the underlying layers results in the lowering of the surface of the drained peatland, which assumes the highest annual rate in the initial 15–20 years after drainage and is called the first subsidence phase [20–26]. The magnitude and rate of peatland surface subsidence depends mainly on the physical properties of the soil, peat type, and drainage depth. Despite the stabilization of the soil water level at a certain reduced level, the rate of the soil surface subsidence decreases but does not cease and passes into the second phase of subsidence [20,22,27–29]. Mainly chemical and biological processes occur in Phase II, resulting in partial (humification) and complete (mineralization) decomposition of organic matter. During these processes, the resulting greenhouse gas, carbon dioxide, is emitted from the atmosphere and dissolved in groundwater [2,7,30–36]. Consequently, this leads to a decrease in organic carbon in the surface peat layers, and an increase in nitrogen, both in the soil and in the water into which it is leached. This leads to upsetting the natural C:N ratio and causing it to narrow, which is a measure of the degradation processes of these soils. As a result of degradation, the mineral content, i.e., ash content, increases [16,29,37–39].

The processes of surface subsidence in drained peat soils, the decrease of their thickness, and even disappearance of these areas from the natural environment have been quite well described in the world literature based on field studies around the world at single measurement points. These phenomena have been monitored over many years at single measurement points, along cross sections, or on the surface of entire peatlands [27,28,40–46]. For example, for drained peat soils located in the Groot-Mijdrecht Polder near Amsterdam (The Netherlands), surface subsidence measurements were made at 1423 locations between 1954 and 1968 [43]. However, less attention has been paid to areas located near drainage channels and ditches where intensification of the surface subsidence process of drained soils is the greatest due to the lowest groundwater table in their vicinity during the drainage process [37,47–49]. In the literature there is much less research related to the effects of described processes on functioning and technical parameters of drainage systems, such as the decreasing depth of ditches and position of drainage pipelines as a result of the subsidence process. Undoubtedly, this creates a threat to the proper functioning of such systems and the necessity of their modernization in terms of conducting proper water management on such areas (irrigation of these areas).

Measurement results for banks and bottoms of 12 ditches located in a drained fen peat are analyzed in this paper and available archival data from the project of this object are used. Based on 1967 (designed) and 2014 (measured) elevation data, the following objectives are undertaken:

1. Determination of the depth of ditches after subsidence (t) in relation to the designed depth (t_0), determination of the amount of their shallowing ($d = t_0 - t$), and the ratio (d/t) in the analyzed period;
2. Determination of the magnitude and average annual rate of subsidence of the peat soil surface near the measurement points along the analyzed ditches;
3. Development of empirical relationships between the amount of subsidence (y) and the rate of average annual surface subsidence of a drained peat deposit (z), and its original (initial) thickness (x) in the vicinity of drainage ditches;
4. Comparison of the results of calculating the amount of subsidence/subsidence rate using the proposed empirical relationships with the results obtained from 10 empirical equations of this type published in the literature.

2. Material and Methods

The study was conducted on a fragment of the Solec fen-peatland (Góra Kalwaria Commune, Piaseczyński Poviat, Masovian Voivodeship, Central Poland (52°2′16.345″ N, 21°6′14.622″ E). The 220-ha peat deposit is made up of quaternary formations up to about 50 m thick. The aquifer consists of sands, and in the upper part there are organic soils, sedge, and sedge-reed peats of medium degree of decomposition [50,51]. In the vicinity of the site boundaries, the peat thickness is approximately 30–40 cm, while in its central part it varies within approximately 130–180 cm, and locally up to 250 cm. The first land draining works at the site were conducted between 1941 and 1943 and consisted mainly of a new section of the Mała River running through the center of the site [52] (Figure 1a). In 1967, a project for a drainage system was created that called for the construction of 62 ditches and about 80 communication and damming structures [52]. Draining works according to the project were carried out in 1967–1971. The site was subdivided into 13 plots in which sub-irrigation system was implemented. The main source of irrigation water was the Mała River and feeder A and R-32 (Figure 1b). Most of the site was in agricultural use as meadows and pastures. Since about 1985, irrigation has no longer been implemented, no damming devices are in place, and the ditches only drain adjacent land, discharging into the Mała River. Most of the area has not been used for agriculture since about 2000.The Mała River drains water in the range of about 100 m on each side. In the growing season the ditches are usually dry, and after short-term intensive rainfall the depth of water in the ditches is about 15–20 cm.

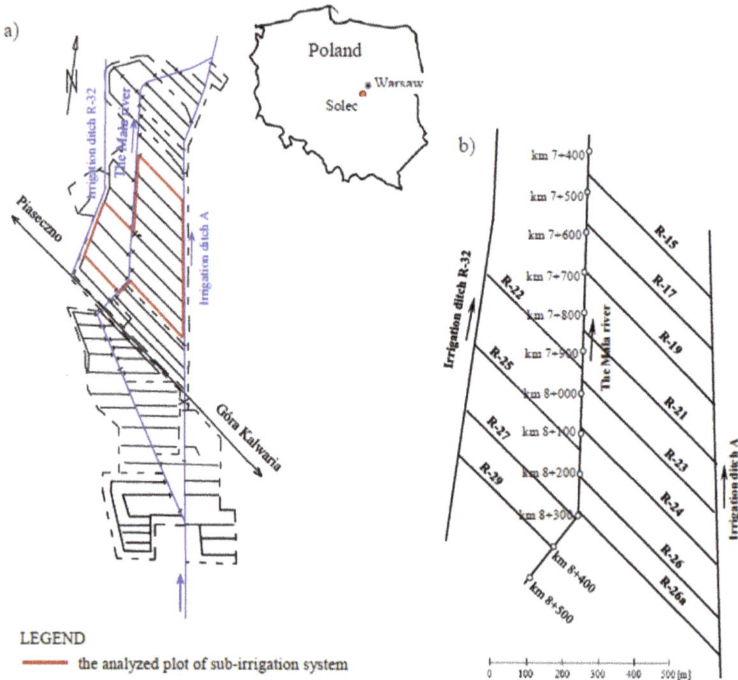

Figure 1. Location of the study site, its schematic (**a**), and the analyzed parts of it (**b**).

The study covered a fragment of the Solec fen with an area of about 50 ha, on which there are 12 drainage ditches: R-15, R-17, R-19, R-21, R-22, R-23, R-24, R-25, R-26, R-26a, R-27, and R-29. Ditch spacing in most cases is 90 m, while ditches R-22 and R-25 are spaced 140 m apart, with lengths ranging from 250 m to about 540 m (Figure 1b). Geodetic measurements of the ordinates of both their banks and bottom were made in cross sections located every 100 m on the ditches in 2014, and the thickness of the peat deposit on both banks was measured. Measurements were taken with a NI 050 levelling device from Carl Zeiss Jena, with reference to the existing state geodetic network.

Changes in bank and bottom ordinates of ditches from 1967 to 2014 were analyzed and ditch depths as designed (t_0) (1967), ditch depths after subsidence (t) (2014), ditch shallowing magnitudes ($d = t_0 - t$), ratio (d/t), and subsidence magnitudes/average annual rate of land surface subsidence near ditches were calculated. The values of the d/t ratio were included in ranges depending on the degree of peat decomposition [53,54]. For peat with a low degree of decomposition (H_1–H_4), the values of this ratio range from 1.28–0.57, for a medium degree of decomposition (H_5–H_6), the ratio (d/t) is 0.39, and for peat with a high degree of decomposition (H_7–H_{10}), it is 0.27–0.19. Based on the results, two empirical relationships were developed between the amount of subsidence (y) (cm) and the rate of average annual subsidence (z) (cm year^{-1}) of the surface of the drained peat deposit and its initial thickness (x) near the drainage ditches. The calculation results obtained using the proposed equations were compared with the calculation results of these quantities obtained using 10 empirical equations developed for the second phase of subsidence of large areas (not only for near-ditch locations) of drained fens published in the literature, summarized in Table 1.

Table 1. Empirical equations for the second phase of peat surface subsidence based on initial peat deposit depth and soil surface subsidence or annual peat subsidence rates published by various authors.

Site (Source)	Equation (Number)		Explanations of Symbols acc. to Sources
Noteć River Valley; drainage intensity of peatlands:			
- low (0.4–0.6 m),	$y = 0.051x + 8.6$	(1)	y—surface subsidence [cm]
- medium (0.6–1.0 m),	$y = 0.05x + 18$	(2)	x—initial peat depth [cm]
- high (1.0–1.2 m),	$y = 0.082x + 34.6$	(3)	
- total (Ilnicki 1972)	$y = 0.101x + 9.5$	(4)	
Peatlands in Central Europe (Ilnicki 1972)	$y = 0.12x + 23$	(5)	y—surface subsidence [cm] x—initial peat depth [cm]
Peatland of Moscow Research Station (Stankiewicz and Karelin 1965 after Ilnicki 1972)	$y = 0.156x + 19.2$	(6)	y—surface subsidence [cm] x—initial peat depth [cm]
Biebrza River Valley; Kuwasy I fen (Krzywonos 1974)	$y = 0.099x + 2.9$	(7)	y—surface subsidence [cm] x—initial peat depth [cm]
Noteć River Valley; drainage intensity of peatlands:			
- medium (0.6–1.0 m),	$z = 0.00107x + 0.34$	(8)	z—average annual rate of soil
- high (1.0–1.2 m),	$z = 0.00228x + 0.47$	(9)	surface subsidence [cm year^{-1}]
- total (Ilnicki 1972)	$z = 0.0021x + 0.17$	(10)	x—initial peat depth [cm]

3. Results

The example longitudinal profile of the selected ditch R-29 (Figure 2) and the cross-sections (Figure 3) show the archival (1967) averaged ordinates of both ditch banks and its bottom, and the results of measurements of the same parameters in 2014, as well as the ordinates of the mineral subsoil. The profiles of 12 ditches included in the drainage system [52] were analyzed. Based on the ordinates of the banks and bottom of the ditches at each hectometer, it was found that the design depth of the ditches (t_o) was within 0.73–1.35 m, with an average of 1.0 m. This indicates a medium-to-high intensity of drainage in the area [55]. The wide range of ditch depths was probably due to the slope of the terrain and the level of the designed bottom. As a result of the dewatering process, it was observed that the depth of the ditches decreased after subsidence (t) in 2014 to an average depth of about 0.74 m, and their shallowing averaged about 0.26 m (Table 2 and Table S1). The values of the ratio of the magnitude of ditch shallowing (d) to its depth after subsidence (t) vary within very large limits from about 0.01 to 1.16, with an average of 0.40. During the analysis period of 47 years (1967–2014), significantly higher values of subsidence were observed on the banks of the ditches compared to the subsidence of their bottoms. The ordinates of the ditch banks decreased by about 0.43 m on average (0.04–0.80 m) compared with the 1967 ordinates, whereas the ditch bottom decreased to a much smaller extent (by about 0.17 m on average), i.e., about 40% in comparison with the amount of their bank subsidence (Table 2 and Table S1). This is also an effect of siltation, the contribution of which is difficult to estimate.

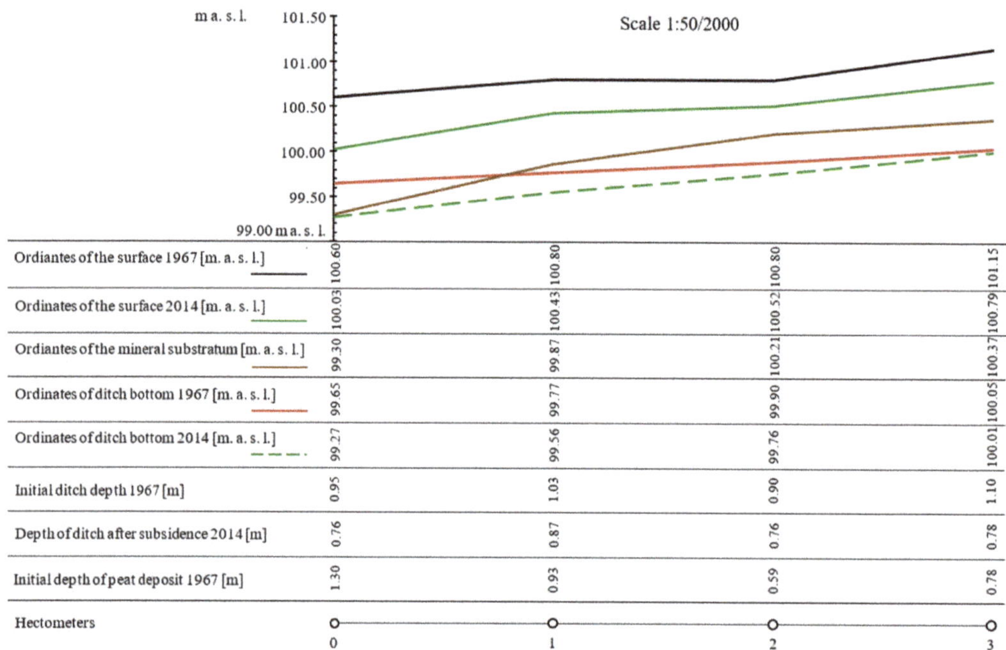

Figure 2. Longitudinal profile of the R-29 ditch.

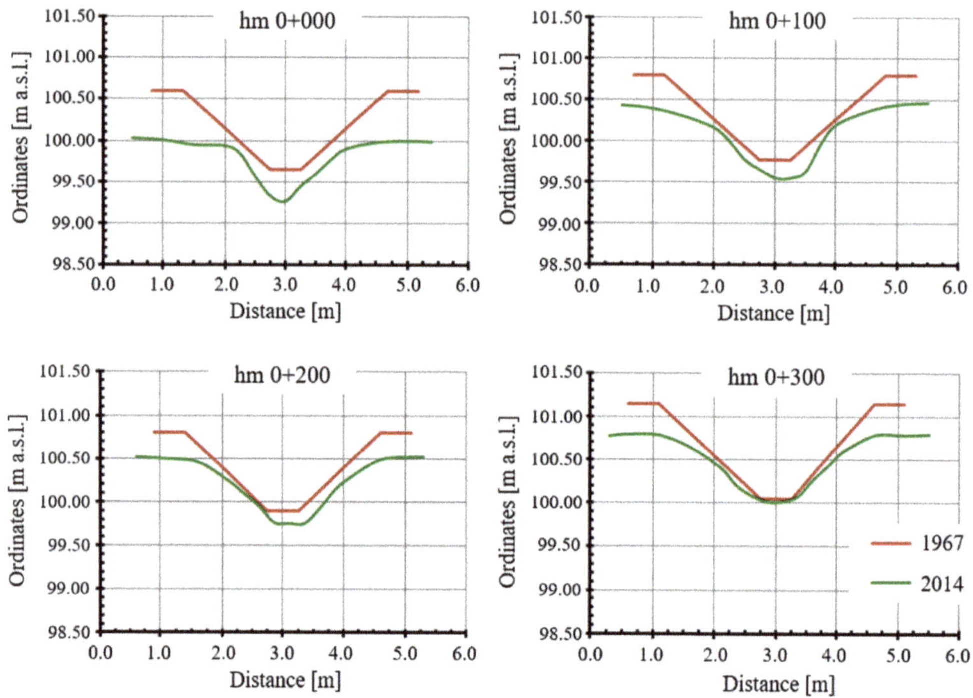

Figure 3. Cross-sections of the R-29 ditch in the year 1967 (designed) and 2014 (measured).

Table 2. Mean values and range (min–max) parameters describing subsidence of drainage ditches (banks and bottom).

Ditch No.	Peat Depth (cm)		Ditch Shallowing (cm)	d/t Ratio (cm)	Subsidence (cm)	
	1967 t_0	2014 t	$d = t_0 - t$		Banks	Bottom
R 15	0.89 (0.73–1.15)	0.55 (0.40–0.75)	0.35 (0.19–0.44)	0.66 (0.35–1.00)	0.63 (0.51–0.70)	0.29 (0.24–0.32)
R 17	0.94 (0.81–1.09)	0.75 (0.63–0.82)	0.20 (0.09–0.34)	0.27 (0.13–0.54)	0.40 (0.26–0.63)	0.20 (0.04–0.36)
R 19	1.10 (1.03–1.15)	0.71 (0.56–0.94)	0.38 (0.20–0.56)	0.59 (0.22–0.95)	0.61 (0.49–0.75)	0.23 (0.00–0.38)
R 21	1.02 (0.93–1.06)	0.71 (0.57–1.00)	0.31 (0.05–0.40)	0.47 (0.05–0.70)	0.59 (0.42–0.80)	0.28 (0.19–0.41)
R 22	1.18 (0.94–1.35)	0.69 (0.50–0.85)	0.49 (0.44–0.53)	0.73 (0.59–0.88)	0.55 (0.49–0.590	0.06 (0.05–0.07)
R 23	0.97 (0.88–1.13)	0.58 (0.42–0.72)	0.39 (0.21–0.51)	0.73 (0.31–1.67)	0.60 (0.42–0.74)	0.21 (0.04–0.40)
R 24	0.98 (0.76–1.12)	0.83 (0.65–1.05)	0.15 (0.01–0.38)	0.20 (0.01–0.58)	0.23 (0.04–0.50)	0.08 (0.03–0.12)
R 25	0.81 (0.73–0.89)	0.72 (0.68–0.77)	0.10 (0.04–0.12)	0.14 (0.06–0.18)	0.21 (0.11–0.29)	0.11 (0.07–0.17)
R 26	0.99 (0.79–1.10)	0.75 (0.54–0.94)	0.23 (0.02–0.46)	0.36 (0.03–0.85)	0.35 (0.06–0.56)	0.11 (0.03–0.21)
R 26a	0.92 (0.85–1.00)	0.71 (0.60–0.85)	0.21 (0.07–0.30)	0.31 (0.09–0.43)	0.34 (0.22–0.44)	0.13 (0.10–0.15)
R 27	0.85 (0.75–0.95)	0.75 (0.61–0.93)	0.10 (0.02–0.16)	0.15 (0.02–0.25)	0.26 (0.07–0.36)	0.16 (0.05–0.27)
R 29	0.98 (0.90–1.10)	0.79 (0.76–0.87)	0.19 (0.03–0.32)	0.24 (0.03–0.41)	0.40 (0.28–0.57)	0.20 (0.04–0.38)
Total	0.97 (0.73–1.35)	0.71 (0.40–1.05)	0.26 (0.01–0.56)	0.41 (0.01–1.17)	0.43 (0.04–0.80)	0.18 (0.00–0.41)

The variability of the initial peat thickness in 1967 along the route of the analyzed ditches was quite large and ranged from about 30 cm to about 255 cm, with an average of about 122 cm (Table 3 and Table S2). The deposit was the shallowest along the route of ditch R-25, and the deepest along ditches R-19 and R-21. As a rule, the highest deposit thicknesses were found in the middle part of the trench route, while the lowest thicknesses were found at the ends of the trenches, near feeders A and R-32.

Table 3. Mean values and range (min–max) of initial peat depth and average rate of subsidence of drainage ditches.

Ditch No.	L (m)	Peat Depth 1967 (cm)	Subsidence y (cm)	Subsidence $y_{(\%)}$ (%)	Average Annual Subsidence Rate z (cm year^{-1})
R 15	450	141.2 (95–255)	63.2 (51–70)	50.3 (27.5–71.6)	1.34 (1.09–1.49)
R 17	470	135.6 (102–200)	40.0 (26–63)	30.6 (17.0–40.4)	0.85 (0.55–1.34)
R 19	490	163.6 (108–226)	61.0 (49–75)	39.2 (23.5–62.8)	1.30 (1.04–1.60)
R 21	500	152.0 (82–237)	58.5 (42–80)	42.7 (23.1–62.8)	1.24 (0.89–1.70)
R 22	260	120.0 (65–165)	55.0 (49–59)	51.8 (34.5–75.4)	1.17 (1.04–1.26)
R 23	500	150.3 (103–223)	60.3 (42–74)	42.3 (33.2–58.6)	1.28 (0.89–1.57)
R 24	510	114.8 (63–160)	22.5 (4–50)	19.0 (4.4–39.1)	0.48 (0.09–1.06)
R 25	310	52.5 (31–95)	21.0 (11–29)	42.6 (30.5–56.8)	0.45 (0.23–0.62)
R 26	510	120.3 (70–170)	35.3 (6–56)	26.9 (8.6–45.0)	0.75 (0.13–1.19)
R 26a	540	109.2 (30–170)	33.8 (22–44)	39.3 (22.4–73.3)	0.72 (0.47–0.94)
R 27	370	72.5 (40–110)	26.0 (7–36)	35.0 (17.5–48.3)	0.55 (0.15–0.77)
R 29	324	90.0 (59–130)	39.5 (28–57)	44.3 (39.8–47.5)	0.84 (0.60–1.21)
Total	-	121.7 (30–255)	43.3 (4–80)	37.7 (4.4–75.4)	0.92 (0.09–1.70)

Explanations: L—total length of the ditch.

Based on the magnitude of bank subsidence of the analyzed 12 ditches over 47 years (1967–2014), it was found that the highest subsidence values were observed in the central part of their route, slightly smaller at the mouth of the ditches feeding into the Mała River, and the lowest near the end of the ditches, which may be related to the variable thickness of the deposit along the route of the ditches (Table 3 and Table S2). The amount of land surface subsidence in the vicinity of the ditches during the analysis period ranged from about 4 cm to 80 cm with an average of 43 cm. This corresponds to a reduction in the thickness of the peat deposit relative to the initial thickness of 4% to 75% (43% on average). The mean annual subsidence rate during the 47-year period analyzed was also variable, ranging from 0.08 to 1.7 cm year^{-1}, with an average of 0.92 cm year^{-1} (Table 3 and Table S2). The results of land surface subsidence measurements from 1967 to 2014 along the analyzed drainage ditches are presented in Figure 4a in relation to the initial thickness of the peat deposit. The average annual subsidence rate of the peat surface depended on the same parameter, as shown in Figure 4b.

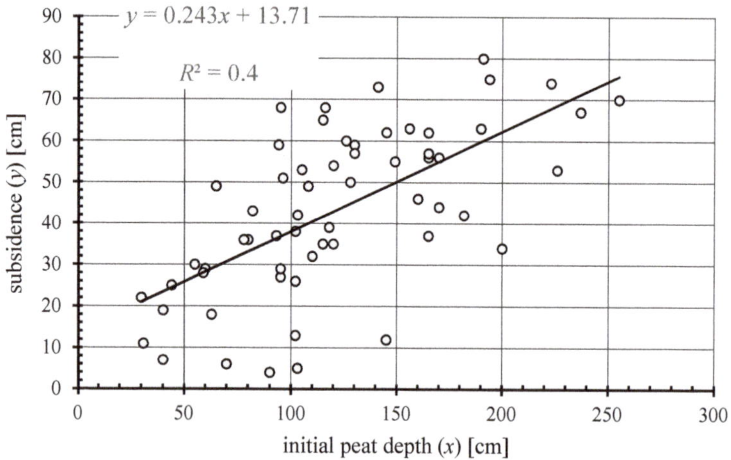

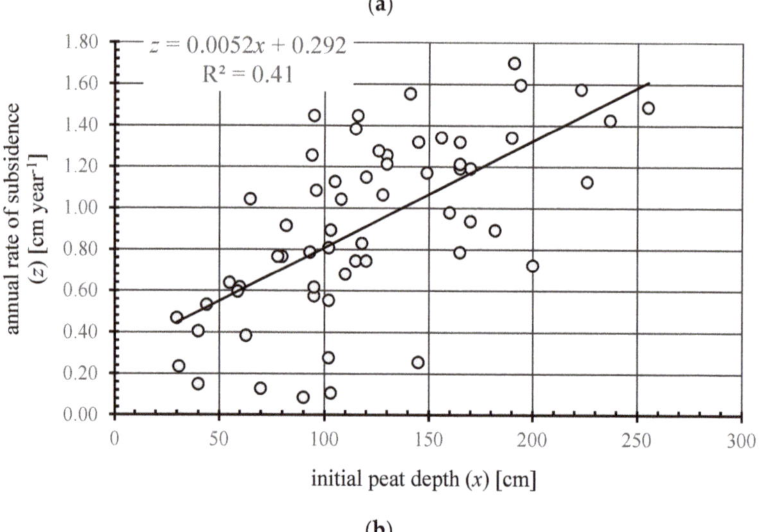

Figure 4. Relationship between the amount of surface subsidence of a drained peat deposit (**a**), and the average annual rate of peat deposit surface subsidence (**b**) on the initial thickness of the peat deposit.

Based on the results obtained (Figure 4a,b), two empirical Equations (11) and (12) were developed, which make the amount of subsidence of the drained peat deposit surface (y) (cm) near the ditches and the average annual rate of peatland surface subsidence (z) (cm year^{-1}) dependent on the initial thickness of the peat deposit (x) (cm) (explanation of symbols and units used in the text):

$$y = 0.243x + 13.71 \tag{11}$$

$$z = 0.0052x + 0.292 \tag{12}$$

The relationships developed are linear, with correlation coefficient r values of 0.65 in both cases. Using the proposed Equations (11) and (12), calculations were performed for the amount of subsidence (y) (Figure 4a) and the average annual subsidence rate (z) (Figure 5b) depending on the initial thickness of the peat deposit (x) within the range of its values from 30 cm to 250 cm (the range of the actual thickness of the deposit in the analyzed fragment of the site). The values of the same parameters (y) and (z) were calculated using Equations (1)–(10) given in Table 1 (Figure 5a,b). The subsidence values calculated according to Equation (11) were larger over the entire range of thicknesses considered than the subsidence volumes calculated by Equations (1), (2), (4), and (7). Equations (1) and (2) were developed for low (0.4–0.6 m) and medium (0.6–1.0 m) drainage intensity conditions, respectively (Table 1). When calculated with Equations (5) and (6), the results coincide with those obtained using Equation (11) in the thickness range from 30 cm to 100 cm. Next, as the thickness increases, the calculation results of the proposed Equation (11) increase significantly with respect to the other equations. In the case of Equation (3) (high drainage intensity, i.e., 1.0–1.2 m) for thickness changes within 30–100 cm, the calculations exceed the settling calculated by Equation (11). In the range of larger thicknesses of the peat deposit (above 100 cm), the results of the subsidence calculation by Equation (3) are similar to those obtained by Equations (5) and (6) (Figure 5a).

In the case of the equations for the average annual subsidence rate (Figure 4b), the calculation of this parameter with the proposed equation (12) outperforms the results of calculations using Equations (8)–(10) (Table 1) in the prevailing range of soil thickness changes, i.e., from 70 cm to 250 cm. Equations (11) and (12) were developed based on the results of settling measurements of the drainage ditch banks, where the intensity of drainage is locally the highest and consequently causes the greatest settling of the soil surface.

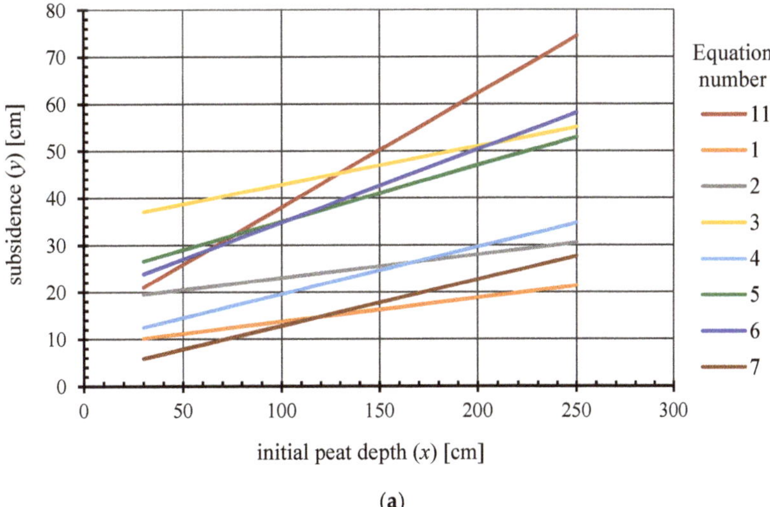

(**a**)

Figure 5. *Cont.*

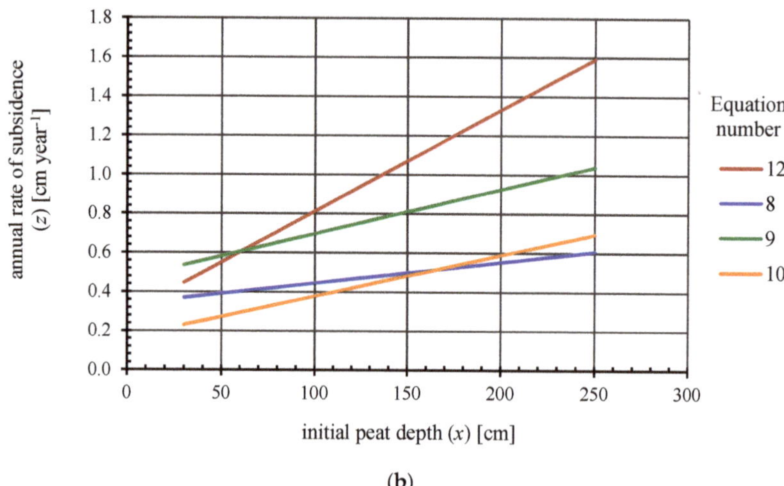

(b)

Figure 5. Comparison of the results of calculating the amount of surface subsidence using proposed Equation (11) (**a**) and the amount of average annual rate of subsidence using proposed Equation (12) (**b**) with the results of calculating these parameters with equations published in the literature.

4. Discussion of Results

The lowering of the water table as a result of dewatering causes the buoyancy force to disappear and the pressure of the dewatered layers on the lower lying layers to disappear, resulting in the subsidence of the peat deposit at depth. The land surface, which is composed of the sum of the subsidence of the individual layers and, to a lesser extent, the bottom of the trenches, decreases the most [22,25,26,28]. When conducting long-term studies (1965–1998) on subsidence of peat soils in northern Poland, it was found that drainage pipelines initially located about 1 m below the ground surface decreased their depth by about 40–50 cm due to surface and bottom subsidence. The average annual rate of decrease in their depth was contained in the range of 1.21–1.51 cm year^{-1}. In the conditions of the northern Netherlands, according to Van den Akker et. al. [26,56], the position of the water table in the ditches on peat soils used as pastures was mostly maintained at about 60 cm below the ground surface. In these soils, there was a decrease in the water table in the ditches and their bottoms observed as a result of the subsidence process, by about 10 cm on average over a period of 10 years. Maintaining the ditch water table between 0.30 and 1.20 m below ground surface on these soils results in an average annual subsidence rate ranging from 0.50 to 2.20 cm year^{-1} [26,56]. Grzywna [57,58], Gąsowska [37], Oleszczuk et al. [49] observed a decrease in the designed ditch depth of 1.00 m on average to a post-subsidence depth of 0.20–0.60 m over a period of about 40–50 years on this type of site (drained peat soils, grassland use) in central and eastern Poland. In the case analyzed in this work, the design depth of the 12 ditches in 1967 ranging from 0.73 to 1.35 m (average depth of about 0.97 m) decreased over a period of 47 years to post-subsidence depths ranging from 0.40 to 1.05 m, with an average of 0.71 m.

Ostromęcki [53] and Pierzgalski [54] determined the ranges for the values of the ditch shallowing post-subsidence depth ratio (*d/t*) in relation to the degree of organic matter decomposition in peat soil. In the studied peatland, the values of this coefficient varied over a very wide range, with an average of 0.41, which indicates that most of these soils had an average degree of organic matter decomposition. This is confirmed by independent physical tests conducted under laboratory conditions on soil samples from this site [37,39,45,50,51].

The value of subsidence of the drained organic soils surface is usually expressed in cm and quite often related as a percentage of its initial thickness. However, average annual rate of subsidence expressed in cm year^{-1} or mm year^{-1} allows the results obtained from this process to be compared across many countries and continents in the world [27].

The average annual subsidence rate in the vicinity of drainage ditches in the study area of Solec peatland was 0.92 cm year^{-1}. Similar studies of the subsidence rates conducted on a section of this peatland by Oleszczuk et al. [45] over a 40-year period (1978–2018) indicated an average annual subsidence rate of 0.62 cm year^{-1}. The authors then showed that the subsidence rate also depended on the location of the measurement points in relation to the existing drainage system. Out of the 14 measurement points analyzed, four of them were located in the immediate vicinity of drainage ditches and the Mała River. The average annual rate was relatively varied across these points and amounted to 0.25 cm year^{-1}, 0.48 cm year^{-1}, 1.20 cm year^{-1}, and 1.45 cm year^{-1}, respectively [45]. It can be concluded that typically the surface settles most along localized drainage facilities (canals, ditches, drains) because the depression curve of the groundwater table near these facilities assumes the greatest depths [37]. This results in a significant decrease in the initial depth of channels and ditches, and a decrease in the depth of drain location. When peatlands are used agriculturally, this poses a threat to the continued proper functioning of drainage systems in these areas [49,59], and requires upgrading such systems by, e.g., dredging the existing ditch network [43,56,60–62]. On the other hand, on sites where agricultural use has already been withdrawn, subsidence and shallowing of this infrastructure may contribute to limiting the functioning of these systems, resulting in a positive effect of inhibiting or blocking water runoff, limiting the degradation of peat soils [62–64]. These topics have been widely reported in the literature. In this context, the authors of this paper emphasize that they focus only on the technical aspect related to the subsidence effects on drainage systems.

The amount of settling of drained peat soils and its rate depends on numerous factors: the depth of drainage of the peat deposit, its initial thickness, the botanical composition of peat and its decomposition degree, physical and chemical properties, the time elapsed since drainage, the type of use, and the climatic conditions in the area [20, 2540]. In order to develop empirical relationships expressing the magnitude/average annual subsidence rate of a drained peat soil surface of practical dimension, it is not possible to consider all these factors. Relationships presented in the literature expressing the magnitude/rate of subsidence have considered the initial thickness of the peat deposit, the length of time since dewatering, the depth/intensity of dewatering, and the position of the groundwater table [20–22,25,27,42,45,58,61,65,66]. Most commonly, simple empirical relationships between the magnitude/rate of annual subsidence and the initial thickness of the peat deposit are found in the literature (see Table 1). Some results of measurements of this type published in the literature show a large variation in the amount of subsidence of peat soils, depending on their initial thickness. For example, in case of an area of fen-peat soils from the Noteć area, Ilnicki [55] showed that with an initial peat deposit thickness of 300 cm, the amount of surface subsidence varied from about 4 cm to about 80 cm over a period of 43 to 66 years. Consequently, the correlation coefficients of this type of linear relationship are $r = 0.56$ for low drainage intensity (0.40 m–0.60 m), $r = 0.25$ for medium drainage intensity (0.60 m–1.00 m), and $r = 0.58$ for high drainage intensity (1.00 m–1.20 m). Based on the results obtained in the peatlands of the Biebrza Marshes, Krzywonos [23] developed a similar linear relationship with a correlation coefficient of $r = 0.624$.

Comparison of the magnitude of subsidence/average annual rate of subsidence over a period of 47 years against the magnitude of the initial thickness of the peat deposit shows a rather large range of field results, which demonstrates the high degree of difficulty in determining this type of relationship. Nevertheless, the correlation coefficients obtained (0.65) are consistent with results published in the literature (e.g., [55]). The analysis of the calculation results for the magnitude and subsidence rate by empirical formulas proposed by different authors indicates that Equations (11) and (12) proposed in this paper can be used to estimate the magnitude of subsidence or the average annual subsidence rate of the drained fen-peat soil surface, with a medium degree of decomposition near drainage ditches in areas with an initial deposit thickness of up to 250 cm.

5. Conclusions

1. Over the period of 47 years (1967–2014) in the analyzed fragment of the Solec fen drainage site near the route of 12 ditches, there were significant changes in the position of their banks and bottoms in relation to the initial values. The effect of these changes reduced the depth of the ditches from their design depths. As a result of the subsidence processes of their banks and bottom, the designed depth of the ditches within the range of 0.73–1.35 m decreased to the values within the range of 0.40–1.05 m. The ditch banks have subsided by about 42 cm on average, and the bottom by about 17 cm on average. As a result, they have been shallowed by about 26 cm on average over a period of 47 years.

2. The ratio of the amount of ditch shallowing to its depth after subsidence (d/t), which is associated with the degree of peat decomposition, was 0.41 in the analyzed peatland section, indicating a medium degree of decomposition. This was confirmed by the results of previous independent, long-term field and laboratory studies at the site.

3. The mean annual subsidence rate of 0.92 cm year^{-1} in the fen-peat soil studied along the route of the drainage ditches was about 48% higher compared to the average of the entire area, confirming the higher intensity of drainage and subsidence of soils near the ditches.

4. The two empirical equations proposed to estimate the amount of subsidence and the average annual subsidence rate of the peatland surface near the route of drainage ditches were applied to a fen with an average degree of organic matter decomposition in area with an initial peat deposit thickness of up to 250 cm.

Supplementary Materials: The following are available online at https://www.mdpi.com/article/10.3390/land10121287/s1, Table S1: Results of field measurements (2014) and archival data (1967) of the analyzed ditches on a part of the Solec peatland; Table S2: Values of initial thickness of the peat deposit (1967) and the size of the subsidence and the average annual rate of subsidence in the years 1967–2014.

Author Contributions: Conceptualization: R.O., E.Z., J.U.; methodology: J.U., R.O., E.Z., validation: R.O., E.Z., J.J.; formal analysis: R.O., E.Z., J.U.; investigation: R.O., J.U.; data curation: R.O., J.U., J.J.; writing—original draft preparation: R.O., E.Z., J.J.; writing review and editing: E.Z., J.J.; visualization: J.U., E.Z.; project administration: E.Z., J.J. All authors have read and agreed to the published version of the manuscript.

Funding: This research received no external funding.

Institutional Review Board Statement: Not applicable.

Informed Consent Statement: Not applicable.

Data Availability Statement: Not applicable.

Conflicts of Interest: The authors declare no conflict of interest.

References

1. Moore, P.D. The future of cool temperate bogs. *Environ. Conserv.* **2002**, *29*, 3–20. [CrossRef]
2. Oleszczuk, R.; Regina, K.; Szajdak, L.; Höper, H.; Maryganowa, V. Impacts of agricultural utilization of peat soils on the greenhouse gas balance. In *Peatlands and Climate Change*; Strack, M., Ed.; International Peat Society: Jyväskylä, Finland, 2008; pp. 70–97.
3. Post, W.M.; Emanuel, W.R.; Zinke, P.J.; Stangenberger, A.G. Soil carbon pools and world life zones. *Nature* **1982**, *298*, 156–159. [CrossRef]
4. Byrne, K.A.; Chonjicki, B.; Christensen, T.R.; Drosler, M.; Freibauer, A.; Friborg, T.; Frolking, S.; Lindroth, A.; Mailhammer, J.; Malmer, N.; et al. EU peatlands: Current carbon stocks and trace gas fluxes. In *Report 4/2004 to 'Concerted Action: Synthesis of the European Greenhouse Gas Budget'*; Geosphere-Biosphere Centre, University of Lund: Lund, Sweden, 2004; p. 58.
5. Jadczyszyn, J.; Bartosiewicz, B. Soil drainage and degradation processes. Studia i Raporty IUNG-PIB. *Puławy* **2020**, *64*, 49–60.
6. Kudlicki, Ł. Desertification threat in Poland. *Bezpieczeństwo Nar.* **2006**, *1*, 201–211.
7. Couwenberg, J. Greenhouse gas emissions from managed peat soils: Is the IPCC reporting guidance realistic? *Mires Peat* **2011**, *8*, 1–10.

8. Okruszko, H. The effect of land reclamation on organic soils within Poland conditions. *Zesz. Probl. Post Nauk. Roln.* **1976**, *177*, 159–204.
9. Okruszko, H.; Ilnicki, P. The moorsh horizons as quality indicators of reclaimed organic soils. In *Organic Soils and Peat Materials for Sustainable Agriculture*; Parent, L.E., Ilnicki, P., Eds.; CRC Press: Boca Raton, FL, USA, 2003; p. 205.
10. Doroszewski, A.; Jóźwicki, T.; Wróblewska, W.; Kozyra, J. *Agricultural drought in Poland in the years 1961–2010*; Institute of Soil Science and Plant Cultivation. State Reaserch Institute: Puławy, Poland, 2014; p. 144.
11. Fenner, N.; Freeman, C. Drought-induced carbon loss in peatlands. *Nature Geosci.* **2011**, *4*, 895–900. [CrossRef]
12. Koza, P.; Łopatka, A.; Jadczyszyn, J.; Wawer, R.; Doroszewski, A.; Siebielec, G. *Raport IUNG-PIB: Delimitation of Areas at Varying Degrees of risk of Drought in Poland for the Purposes of Implementing the Operation "Modernization of Agricultural Farms" in the Rural Development Program for 2014–2020*; IUNG: Puławy, Poland, 2019; p. 43.
13. Brandyk, T.; Szatyłowicz, J.; Oleszczuk, R.; Gnatowski, T. Water–related physical attributes of organic soils. In *Organic Soils and Peat Materials for Sustainable, Agriculture*; Parent, L.E., Ilnicki, P., Eds.; CRC Press: Boca Raton, FL, USA, 2003; pp. 33–66.
14. Gebhard, S.; Fleige, H.; Horn, R. Anisotropic shrinkage of mineral and organic soils and its impact on soil hydraulic properties. *Soil Till. Res.* **2012**, *125*, 96–104. [CrossRef]
15. Hooijer, A.; Page, S.; Jauhiainen, J.; Lee, W.A.; Lu, X.X.; Idris, A.; Anshari, G. Subsidence and carbon loss in drained tropical peatlands. *Biogeosciences* **2012**, *9*, 1053–1071. [CrossRef]
16. Okruszko, H. Transformation of fen-peat soils under the impact of draining. *Zesz. Probl. Post. Nauk. Roln.* **1993**, *40*, 3–73.
17. Peng, X.; Horn, R. Anisotropic shrinkage and swelling of some organic and inorganic soils. *Eur. J. Soil Sci.* **2007**, *58*, 98–107. [CrossRef]
18. Silins, U.; Rothwell, R.L. Forest peatland drainage and subsidence affect soil water retention and transport properties in an Alberta peatland. *Soil Sci. Soc. Am. J.* **1998**, *62*, 1048–1056. [CrossRef]
19. Snyder, G.H. Everglades agricultural area soil subsidence and land use projections. *Annu. Proc. Soil Crop Sci. Soc. Fla.* **2005**, *64*, 44–51.
20. Eggelsmann, R. Subsidence of peatland caused by drainage, evaporation and oxidation. In *Land Subsidence, Proceedings of the Third International Symposium on Land Subsidence Held in Venice, Italy, 19–25 March 1984*; IAHS Publication No. 151; Johnson, A.I., Carbognin, L., Ubertini, L., Eds.; Institute of Hydrology, Wallingford: Oxfordshire, UK, 1986; pp. 497–505.
21. Hillman, G.H. Effects of engineered drainage on water tables and peat subsidence in an Alberta Treed Fen. In *Northern Forested Wetlands Ecology and Management*; Carl, C., Ed.; CRC Lewis Publishers: Boca Raton, FL, USA, 1997; pp. 253–272.
22. Jurczuk, S. The influence of water regulations on subsidence and mineralisation of organic soils. In *Biblioteczka Wiadomości IMUZ*; IMUZ: Falenty, Poland, 2000; Volume 96, p. 120.
23. Krzywonos, K. The subsidence of peatlands after drainage near ZD IMUZ Biebrza. *Wiadomości IMUZ Falenty* **1974**, *12*, 151–169.
24. Lipka, K.; Zając, E.; Wdowik, W. The effect of land use on the disappearance of peat-moorsh soils in the Mrowla River valley near Rzeszów. *Zesz Probl Post Nauk Roln* **2005**, *507*, 349–355.
25. Schothorst, C.J. Subsidence of low moor peat soils in the western Netherlands. *Geoderma* **1977**, *17*, 265–291. [CrossRef]
26. Wösten, J.H.M.; Ismail, A.B.; van Wijk, A.L.M. Peat subsidence and its practical implications: A case study in Malaysia. *Geoderma* **1997**, *78*, 25–36. [CrossRef]
27. Ikkala, L.; Ronkanen, A.-K.; Utriainen, O.; Kløve, B.; Marttila, H. Peatland subsidence enhances cultivated lowland flood risk. *Soil Till Res* **2021**, *212*, 105078. [CrossRef]
28. Lipka, K.; Zając, E.; Hlotov, V.; Siejka, Z. Disappearance rate of a peatland in Dublany near Lviv (Ukraine) drained in 19th century. *Mires Peat* **2017**, *19*, 1–15.
29. Maslov, B.S.; Kolganov, A.V.; Kreshtapova, V.N. *Peat Soils and Their Change under Amelioration*; Rossel'khozizdat: Moskva, Russia, 1996.
30. Berglund, Ö.; Berglund, K. Distribution and cultivation intensity of agricultural peat and gyttja soils in Sweden and estimation of greenhouse gas emissions from cultivated peat soils. *Geoderma* **2010**, *154*, 173–180. [CrossRef]
31. Camporese, M.; Gamolati, G.; Putti, M.; Teatini, P. Peatland Susidence in the Venice Watershed. In *Peatlands: Evolution and Records of Environmental and Climate Changes*; Martini, I.P., Martinez Cortizas, A., Chesworth, W., Eds.; Elsevier: Amsterdam, The Netherlands, 2006; pp. 529–550.
32. Ilnicki, P. *Restoration of Carbon Sequestrating Capacity and Biodiversity in Abandoned Grassland on Peatland in Poland. Monography*; August Cieszkowski Agricultural University Press of Poznań: Poznań, Poland, 2002; p. 170.
33. Kasimir-Klemedtsson, A.; Klemedtsson, L.; Berglund, K.; Martikainen, P.; Silvola, J.; Oenema, O. Greenhouse gas emissions from farmed organic soils: A review. *Soil Use Manag.* **1997**, *13*, 245–250. [CrossRef]
34. Kechavarzi, C.; Dawson, Q.; Leeds-Harrison, P.B.; Szatyłowicz, J.; Gnatowski, T. Water–table management in lowland UK peat soils and its potential impact on CO_2 emission. *Soil Use Manag.* **2007**, *23*, 359–367. [CrossRef]
35. Schipper, L.A.; McLeod, M. Subsidence rates and carbon loss in peat soils following conversion to pasture in the Waikato region, New Zealand. *Soil Use Manag.* **2002**, *18*, 91–93. [CrossRef]
36. Van den Akker, J.J.H.; Jansen, P.C.; Hendriks, R.F.A.; Hoving, I.; Pleijter, M. Submerged infiltration to halve subsidence and GHG emissions of agricultural peat soils. In Proceedings of the 14th International Peat Congress, Extended abstract No. 383, Stockholm, Sweden, 3–8 June 2012; p. 6.

37. Gąsowska, M. The Influence of Change in Meadow Use on Physical–Water Properties of Organic Soils. Case Study: Łąki Soleckie. Ph.D. Thesis, Warsaw University of Life Sciences–SGGW, Warsaw, Poland, 2017.
38. *Organic Soils and Peat Materials for Sustainable Agriculture*; Parent, L.E.; Ilnicki, P. (Eds.) CRC Press: Boca Raton, FL, USA, 2003; p. 205.
39. Truba, M.; Oleszczuk, R. An analysis of some basic chemical and physical properties of drained fen peat and moorsh soil layers. *Ann Wars. Univ. Life Sci.-SGGW Land Reclam.* **2014**, *46*, 69–78. [CrossRef]
40. Dawson, Q.; Kechavarzi, C.; Leeds-Harrison, P.B.; Burton, R.G.O. Subsidence and degradation of agricultural peatlands in the Fenlands of Norfolk, UK. *Geoderma* **2010**, *154*, 181–187. [CrossRef]
41. Deverel, S.J.; Ingrum, T.; Leighton, D. Present-day oxidative subsidence of organic soils and mitigation in the Sacramento-San Joaquin Delta, California, USA. *Hydrogeol. J.* **2016**, *24*, 569–586. [CrossRef] [PubMed]
42. Evans, C.D.; Williamson, J.M.; Kacaribu, F.; Irawan, D.; Suardiwerianto, Y.; Hidayat, M.F.; Laurén, A.; Page, S.E. Rates and spatial variability of peat subsidence in Acacia plantation and forest landscapes in Sumatra, Indonesia. *Geoderma* **2019**, *338*, 410–421. [CrossRef]
43. Hoogland, T.; van den Akker, J.J.H.; Brus, D.J. Modeling the subsidence of peat soils in the Dutch coastal area. *Geoderma* **2012**, *171–172*, 92–97. [CrossRef]
44. Ilnicki, P.; Szajdak, L.W. *Peatland Disappearance*; PTPN Press: Poznań, Poland, 2016; p. 312.
45. Oleszczuk, R.; Zając, E.; Urbański, J. Verification of empirical equations describing subsidence rate of peatland in Central Poland. *Wetl. Ecol Manag.* **2020**, *2*, 495–507. [CrossRef]
46. Zanello, F.; Teatini, P.; Putti, M.; Gambolati, J. Long term peatland subsidence: Experimental study and modeling scenarios in the Venice coastland. *J. Geophys. Res.* **2011**, *116*, 1–14. [CrossRef]
47. Ciepielowski, A.; Włodarczyk, A. Hydrotechnical structures in protected areas. *Acta Sci. Pol. Arch.* **2004**, *3*, 3–21.
48. Murat-Błażejewska, S.; Zbierska, J.; Ławniczak, A.; Kanclerz, J.; Kupiec, J.; Sojka, M. Exploitation of water infrastructure in aspect of water resources of lowland river catchment. *Acta Sci. Pol. Arch.* **2008**, *7*, 13–22.
49. Oleszczuk, R.; Gąsowska, M.; Guz, G.; Urbański, J.; Hewelke, E. The influence of subsidence and disappearance of organic moorsh soils on longitudinal sub-irrigation ditch profiles. *Acta Sci Pol. Cir.* **2017**, *16*, 3–13. [CrossRef]
50. Brandyk, T. Principles of Moisture Management for Shallow Water Table Soils. In *Treatises and Monographs*; Warsaw Agricultural University Press: Warsaw, Poland, 1990; Volume 116, p. 120.
51. Kaca, E. Mathematical Model of Ground Water Rising in Sub-Irrigation System. Ph.D. Thesis, Institute of Land Reclamation and Water Management, Warsaw Agricultural Unviersity, Warsaw, Poland, 1981; p. 192.
52. Brożek, W. *The Description of the Technical Project of Land Reclamation of the Grassland Mała River Site*; CBSiPWM: Warsaw, Poland, 1967; p. 45, unpublished work.
53. Ostromęcki, J. The Drainage of Grasslands. Guide–Book. Warsaw Agricultural University Press: Warsaw, Poland, 1971; p. 163.
54. Pierzgalski, E. *Land Reclamation of Grassland–Sub-Irrigation Systems. Guide–Book*; Warsaw Agricultural University Press: Warsaw, Poland, 1996; p. 200.
55. Ilnicki, P. Subsidence of Fenland Areas in the Noteć Valley under Longterm Agricultural Use Depending on Their Structure and Drainage Intensity. Doctoral Dissertation, Agricultural University, Szczecin, Poland, 1972; p. 63.
56. Van den Akker, J.J.H.; Hendriks, R.F.A.; Pleijter, M. CO_2 emissions from peat soils in agricultural use: Calculation and prevention. *Agrociencia* **2012**, *16*, 43–50.
57. Grzywna, A. Peatlands surface subsidence in drainage grassland Polesie Lubelskie. *Acta Sci. Pol. Cir.* **2016**, *15*, 81–89. [CrossRef]
58. Grzywna, A. The degree of peatland subsidence resulting from drainage of land. *Environ. Earth Sci.* **2017**, *76*, 1–8. [CrossRef]
59. Brandyk, A.; Oleszczuk, R.; Urbański, J. Estimation of organic soils subsidence in the vicinity of hydraulic structures—Case study of a subirrigation system in Central Poland. *J. Ecol. Eng.* **2020**, *21*, 64–74. [CrossRef]
60. Querner, E.P.; Jansen, P.C.; van den Akker, J.J.H.; Kwakernaak, C. Analysing water level strategies to reduce soil subsidence in Dutch peat meadows. *J. Hydrol.* **2012**, *446–447*, 59–69. [CrossRef]
61. Van den Hardeveld, H.A.; Driessen, P.P.J.; Schot, P.P.; Wassen, M.J. An integrated modelling framework to assess long-term impacts of water management strategies steering soil subsidence in peatlands. *Environ. Impact Asses. Rev.* **2017**, *66*, 66–77. [CrossRef]
62. Van den Hardeveld, H.A.; de Jong, H.; Knepfle, M.; de Lange, T.; Schot, P.P.; Spanjers, B.; Teurlincx, S. Integrated impact assessment of adaptive management strategies in a Dutch peatland polder. In *Proceedings of IAHS Proceedings of the International Association of Hydrological Sciences (PIAHS), Copernicus Publications, Gottingen, Germany, 2020*; Volume 382, pp. 553–557.
63. Clark, D.; Rieley, J. (Eds.) *Strategy for Responsible Peatland Management*; International Peat Society: Jyväskylä, Finland, 2010; p. 39.
64. Joosten, H.; Clarke, D. *Wise Use of Mires and Peatlands-Background and Principles Including a Framework for Decision-Making*; International Mire Conservation Group and International Peat Society: Saarijärvi, Finland, 2002; p. 304.
65. Urzainki, I.; Laurén, A.; Palviainen, M.; Haahti, K.; Budiman, A.; Basuki, I.; Netzer, M.; Hökkä, H. Canal blocking optimization in restoration of drained peatlands. *Biogeosciences* **2020**, *17*, 4769–4784. [CrossRef]
66. Gąsowska, M.; Oleszczuk, R.; Urbański, J. The estimation of the subsidence rate of drained peatland and verification of empirical equations of this process. *Sci. Rev.—Eng. Environ. Sci.* **2019**, *28*, 95–104.

 *land*

Article

Impact of Land Use Changes on the Diversity and Conservation Status of the Vegetation of Mountain Grasslands (Polish Carpathians)

Jan Zarzycki [1,*], Joanna Korzeniak [2] and Joanna Perzanowska [2]

1   Faculty of Environmental Engineering and Land Surveying, University of Agriculture in Krakow, al. Mickiewicza 21, 31-120 Kraków, Poland
2   Institute of Nature Conservation, Polish Academy of Sciences, al. Mickiewicza 33, 31-120 Kraków, Poland; korzeniak@iop.krakow.pl (J.K.); perzanowska@iop.krakow.pl (J.P.)
*   Correspondence: jan.zarzycki@urk.edu.pl

Abstract: In recent decades, in the Polish Carpathians, agriculture has undergone major changes. Our goal was to investigate whether the former management (plowing or mowing and grazing) had an impact on the current species composition, diversity and conservation status of the vegetation of grazing areas. We carried out vegetation studies on 45 grazing sites with traditional methods of grazing (transhumant pastoralism). The survey covered both old (continuous) grasslands and grasslands on former arable land. The most widespread were *Cynosurion* pastures and mesic *Arrhenatherion* grasslands. Wet *Calthion* meadows occurred at more than a half of grazing sites, while nutrient-poor *Nardetalia* grasslands were only recorded at several grazing sites. For each grazing site, we used soil maps from the 1960s to read land use in the past. We mapped present grassland and arable land area. Compared with the 1960s, there was a significant decrease in the area of arable land and an increase in grasslands. Species diversity was greater in grazing sites where grasslands developed on former arable land. However, this diversity was associated mainly with the occurrence of common grassland species. *Cynosurion* pastures and wet *Calthion* meadows had the best conservation status, while nutrient-poor *Nardetalia* grasslands were the worst preserved. We concluded that the conservation status of mesic grasslands and pastures is dependent on the present diversity of land use within a grazing site, rather than the land use history 60 years ago. This is the first study of the natural, not economic, value of pasture vegetation in the Polish part of the Carpathians.

Keywords: grazing management; biodiversity; high-nature-value farming; old field grassland

Citation: Zarzycki, J.; Korzeniak, J.; Perzanowska, J. Impact of Land Use Changes on the Diversity and Conservation Status of the Vegetation of Mountain Grasslands (Polish Carpathians). *Land* **2022**, *11*, 252. https://doi.org/10.3390/land11020252

Academic Editor: Alexandru-Ionuţ Petrişor

Received: 13 December 2021
Accepted: 1 February 2022
Published: 8 February 2022

Publisher's Note: MDPI stays neutral with regard to jurisdictional claims in published maps and institutional affiliations.

Copyright: © 2022 by the authors. Licensee MDPI, Basel, Switzerland. This article is an open access article distributed under the terms and conditions of the Creative Commons Attribution (CC BY) license (https://creativecommons.org/licenses/by/4.0/).

1. Introduction

Grasslands provide a variety of ecosystem services, such as cultural landscape and environmental values [1]. They are significant reservoirs of biodiversity and belong to the most species-rich plant communities in Europe [2]. Apart from an abundance of plant species, they are also the habitat for a multitude of animal species from different taxonomic groups, such as insects, arachnids, and snails [3]. The biodiversity of grassland is also related to the time since the transformation of arable land into grassland.

The majority of grasslands in Europe originated as a result of the interaction of environmental conditions and farming by grazing and mowing [4]. However, social and economic transformations in the 20th century provoked changes in land use and management practices in agriculture. In Western Europe, this process particularly accelerated in the 1950s/1960s [5], while in Eastern Europe after the political transformation at the beginning of the 1990s [6]. As a result, the area covered by grasslands has been decreasing. On the one hand, wherever land is suitable for agricultural use, grasslands are transformed into arable fields or the intensity of such use increases [7,8], which results in the development of highly productive but species-poor communities [9]. On the other hand, in marginal areas less suitable for agriculture, land use is abandoned, which promotes secondary forest succession and the disappearance of grassland communities [10–12]. It is particularly disadvantageous because grassland communities in infertile habitats are characterized by the highest natural value and diversity [9,13]. In the Carpathians, the expansion of forest cover has been progressing for over a century and has especially hastened in recent decades [14]. Despite a significant increase in forest cover, the area of grasslands is not only not reduced but even grows [15], as former arable fields are transformed into meadows and pastures. Similar processes were also observed in other montane regions, e.g., in the Caucasus [16], in Slovakia [17] and in the Apennine mountains [18]. As a consequence, a large portion of grasslands in the Carpathians is situated at present on the former arable fields. They are usually located at lower altitudes and on more fertile soils. The species composition of such communities also formed over a short period. For this reason, these plant communities are usually species-poor and are characterized by a lack of many specialist grassland species [19,20].

A number of studies indicate that the species composition of communities, their diversity and the occurrence of rare species are related not only to the current management but also to the habitat history [21,22]. The historical land use and land use sequences shaped the vegetation of Swedish semi-natural grassland more so than the current management. The highest diversities of grassland plants have been found in pastures continuously grazed since the 18th century [21]. The continuity of extensive mowing and grazing has a positive effect on the occurrence of grassland and forest edge specialist species [22]. The spatial context is also important: the present and historical habitat connectivity influences fine-scale plant species diversity in grazed temperate semi-natural grasslands [23].

Traditional pastoral systems in many mountain areas (Carpathians, Alps, Pyrenees) consisted of transhumance, i.e., grazing livestock at higher altitudes in summer and descending with the herd into valleys for the winter [24]. Such a pastoral system has been used up till now in Romania [25], Ukraine [26], Switzerland, Scandinavia, France and Spain [27]. In Poland, at the end of the 1980s, the period of transition into an open-market economy saw a livestock population crash; for instance, the sheep population was dramatically reduced from 4,837,000 in 1985 to 266,911 in 2018 [28], due to the decreased profitability of agricultural production. However, since the mid-2000s, we have witnessed a revival of transhumant pastoralism in the Polish Carpathians [29] thanks to subsidies under the Common Agricultural Policy (CAP). However, the most common locations of transhumance have changed [29]. New grazing sites are located at lower altitudes and primarily on the site of arable fields (Figure 1). The term grazing sites suggests that they are covered only by pastures. However, in practice, parcels of land used as grazing sites are very diverse and harbor natural and semi-natural types of vegetation, and also arable fields. A sheep herd migrates daily within the grazing sites and in the evening the sheep return to a portable shelter. Biodiversity conservation in rural areas is analyzed in connection with viewing these areas as a patch–corridor matrix, which is a dynamic system linking natural processes with current human impact and its future consequences [30]. This approach, based on a landscape-level perspective, is increasingly often adopted in the conservation of semi-natural grasslands [31,32].

The aim of this work was to characterize the sites used for traditional pastoralism, treated as a kind of functional unit, in terms of (a) the species composition and diversity of plant communities occurring in these areas, (b) the evaluation conservation status of these communities, and (c) the impact of land use in the past and historical and present grassland area to arable land area ratio, on the species diversity and conservation status of grassland.

Figure 1. Grazing on former arable land in the Pieniny Mts.

2. Materials and Methods

2.1. Study Area

The study was carried out at 45 grazing sites located in 7 physico-geographical regions in the Polish part of the Western Carpathians (Figure 2) extending 45 km N–S and 95 km E–W. The extent of the area from the west to the east in combination with a wide range of altitudes (347–1159 m a.s.l.) caused large variations in climatic conditions. The area of grazing sites was diverse and ranged from 15 to 145 hectares (Table 1). All study sites have been extensively grazed by sheep (livestock density did not exceed 0.5 LU/ha) for at least a few years, but most of them have been used in this way for a long time. The survey covered both old (continuous) grasslands and those on former arable land. Vegetation of the grazing sites primarily comprised different types of grasslands with scattered patches of forest, thickets, fields and fallow lands, wetlands covered by fen and marsh vegetation, and sporadic small orchards.

2.2. Methods

Vegetation studies were performed in summer 2017. To analyze the species composition of the main types of grasslands, phytosociological relevés were taken according to the Braun-Blanquet method in 25 m^2 plots [33]. Since the grazing sites differed in area, we took 1 relevé per 10 ha of grazing site area. In total, in all 45 sites, 420 relevés were performed. The thickness of the litter layer was also measured for each relevé. In order to assess the conservation status of the grassland types within each grazing site, their structure and function were evaluated (a proxy of habitat quality). The assessment was based on the methodology applied in the Monitoring of Natural Habitats in Natura 2000 areas, which is conducted in Poland by the Chief Inspectorate for Environmental Protection [34]. Since many patches of mesic grasslands were of transitional nature between habitat types 6510 (lowland hay meadows) and 6520 (mountain hay meadows), they were analyzed jointly as mesic *Arrhenatherion* grassland. Nutrient-poor *Nardetalia* grasslands were classified as the Natura 2000 habitat 6230. The assessment of the structure and function of *Cynosurion* pastures and wet *Calthion* meadows, not included in Annex I to the Habitats Directive [35], was based on an analogical formula, as for habitats 6230, 6510 and 6520. Habitat structure and function were assessed according to a three-point scale: favorable, unfavorable inadequate and unfavorable bad. The composite result of the evaluation of indices was defined individually for each habitat. The following indices were assessed: spatial structure of habitat patches, diagnostic species, invasive alien species, expansive species of herbs, and expansion of shrubs and trees [34]. The nomenclature of vascular plant species was adopted according to Mirek et al. [36]. Field work also included mapping the grassland and arable land area.

The variability of the species composition of grassland vegetation was assessed based on phytosociological relevés. To reduce redundancy due to the oversampling of some areas, the set of 420 relevés was subjected to stratified resampling [37] in JUICE 7.0 [38]. We assumed that the mean geographical distance between all pairs of relevés belonging to the same grassland type should be at least 0.5 km, which corresponds to a geographical grid of 0.25 latitude × 0.4 longitude. The resulting data set used in ordination analyses contained 294 relevés. The unweighted mean Ellenberg Indicator Values (EIV) were calculated for every relevé.

Agricultural soil maps from the 1960s, digitized and georeferenced using QGIS 3.16.7., were used to calculate the area of grassland and arable land at that time in each grazing site. As a quality assessment of grazing sites for agricultural production, indicators of the quality and usefulness of agricultural soils (QUAS) were used [39]. These indicators are the result of calculations made for soil valuation classes and soil-agricultural complexes shown on soil maps. The method of land use (arable vs. grassland) where the relevé was located was also read from the soil maps.

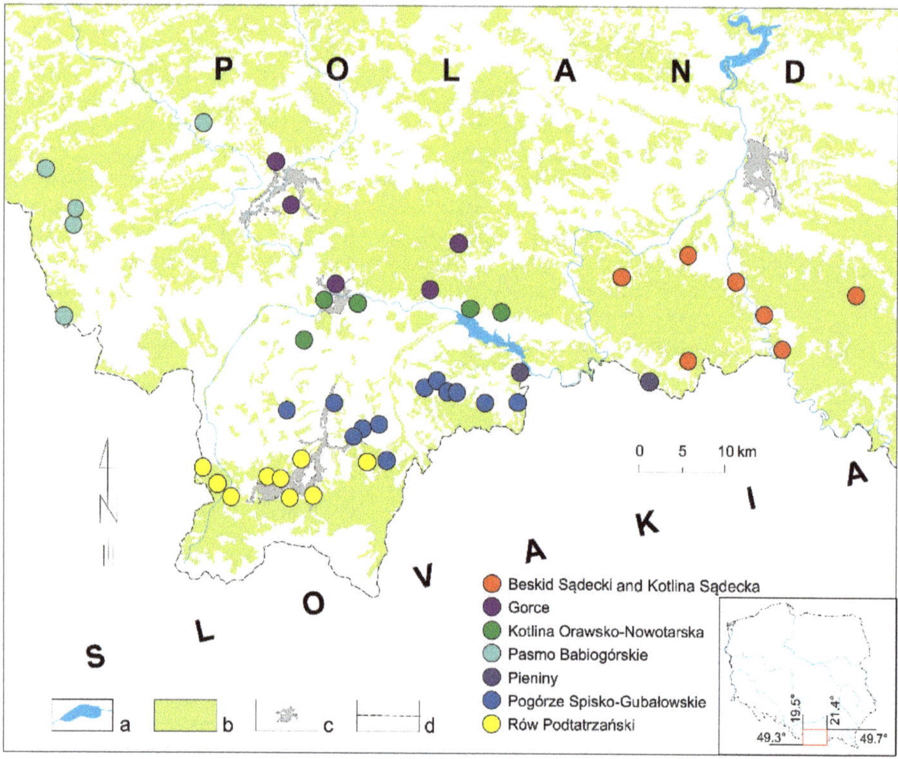

Figure 2. Locations of the grazing sites within regions in the Polish Carpathians. The regions are displayed as colored circles. Explanations: (**a**) rivers and water bodies, (**b**) forests, (**c**) localities, (**d**) state border.

Table 1. Diversity of grassland types in relation to land use in the past. Land use in the 1960s: G-grassland; A-arable fields.

Region (No of Grazing Sites)	Altitude Range (m a.s.l.)	Mean Slope Angle (deg)	Annual Mean Temperature (°C)	Annual Mean Precipitation (mm)
Pasmo Babiogórskie (5)	481–1037	6.5	6–7	830–980
Kotlina Orawsko-Nowotarska (5)	567–807	4.4	6–7	730–850
Pogórze Spisko-Gubałowskie (12)	588–1055	9.2	6–8	540–930
Rów Podtatrzański (9)	794–1159	9.3	4–5	1000–1080
Gorce (5)	518–1012	10.4	4–7	820–1350
Pieniny (2)	490–924	11.5	6–7	730–760
Beskid Sądecki (6) and Kotlina Sądecka (1)	347–908	11.9	5–8	700–850

The diagnostic species [40] for the grassland phytosociological syntaxa (*Calthion, Arrhenatherion, Polygono-Trisetion, Cynosurion, Nardetalia, Arrhenatheretalia, Molinio-Arrhenathereatea*) were distinguished and used to classify relevés into the main types of grassland vegetation. The main diversity measures for sites were determined: average number of vascular plant species in a relevé (α diversity), total number of vascular plant species in all relevés within a grazing site (γ diversity) and the number of diagnostic species for grassland syntaxa.

To explore major gradients in species composition of the main grassland types and their relationship with environmental characteristics and past land use, the Detrended Correspondence Analysis (DCA) was performed using the CANOCO 5.10. For the ordination analysis, species abundance data from 294 plots were square-root transformed with the downweighting of rare species [41]. Because of a lack of normality, the Spearman rank-correlation coefficient for the analyses of relationships between site species diversity, habitat quality and site environmental characteristics was used. Correlations were analyzed using Statistica 13.1.

3. Results

3.1. Changes in Land Use

In the 1960s, arable fields dominated in the areas of analyzed grazing sites, and only 3 sites were entirely without them (Figure 3). In 8 sites, 90% of the area was covered by arable fields. In 2017, the area of arable land was marginal and 22 grazing sites were completely devoid of them, while in 16, the share of arable fields to the total grazing site area was lower than 10%. Only in two grazing sites was the share of arable fields relatively high, amounting to 40%. A QUAS calculated for grazing sites were low and as many as 39 grazing sites did not exceed 35 points (on an 18–100-point scale) (Figure 4).

3.2. Species Composition and Diversity of Four Types of Grasslands

The most widespread vegetation types were pastures of the *Cynosurion* R. Tx. 1947 alliance, mostly *Festuco-Cynosuretum* Büker 1941, and mown and grazed grasslands of the *Arrhenatherion elatioris* (Br.-Bl. 1925) Koch 1926 alliance. Due to a diverse pattern of land use as hay meadows and pastures, many phytocenoses were of a transitional nature, between a pasture and a hay meadow. Wet eutrophic meadows of the *Calthion palustris* R. Tx. 1936 em. Oberd. 1957 alliance occurred at more than a half of grazing sites. *Cirsietum rivularis* Nowiński 1927 was encountered most often, *Scirpetum sylvatici* Ralski 1931 was rarer while *Epilobio-Juncetum effusi* Oberd. 1957 occurred at intensely grazed places. Nutrient-poor grasslands with *Nardus stricta* of the *Nardetalia* Prsg. 1949 order were only recorded at 18 grazing sites. They usually appeared as small patches, often with a large proportion of species typical of the neighboring meadows.

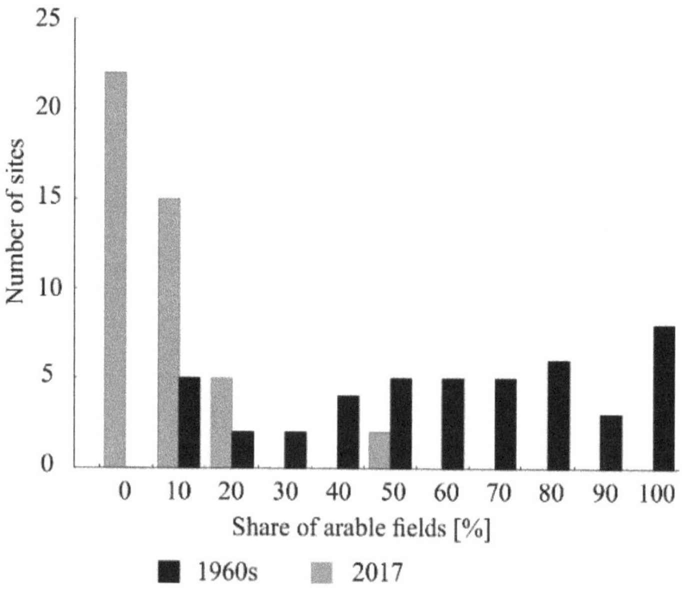

Figure 3. Number of grazing sites in classes of arable field area in the 1960s and in 2017.

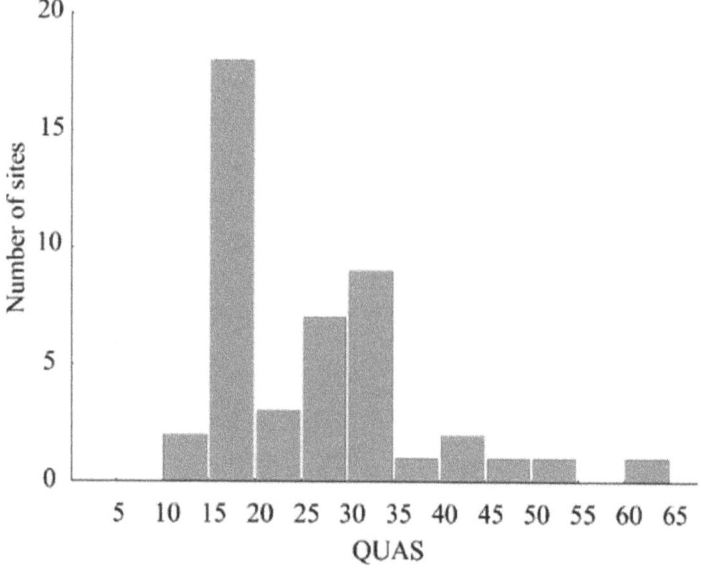

Figure 4. Number of grazing sites in classes of quality and usefulness of agricultural soils index (QUAS).

The main grassland vegetation variability gradient represented by the first DCA axis is associated with soil moisture (Figure 5). The second DCA axis mostly reflects differences in soil fertility, separating nutrient-poor *Nardetalia* grasslands from the remaining grassland types. The occurrence of nutrient-poor grasslands is positively correlated with slope and litter thickness, which suggests a lack of agricultural use. Wet *Calthion* meadows are associated with "old" grasslands. For the remaining types of grasslands, historical land use either as arable fields or grassland is not important. The diagram shows a large degree of similarity between the species composition of *Cynosurion* pastures and mesic *Arrhenatherion* grasslands, and dissimilarity with wet meadows and, to a lesser extent, also *Nardetalia* grasslands. Wet *Caltion* meadows were observed to include a lower proportion of the relevés located on former arable fields compared with the remaining grassland types (Table 2). On the other hand, relevés classified as mesic *Arrhenatherion* grasslands occur more often on former arable fields than on "old" grasslands. In pastures and nutrient-poor *Nardetalia* grasslands, the number of relevés taken on former fields and "old" grasslands is almost equal.

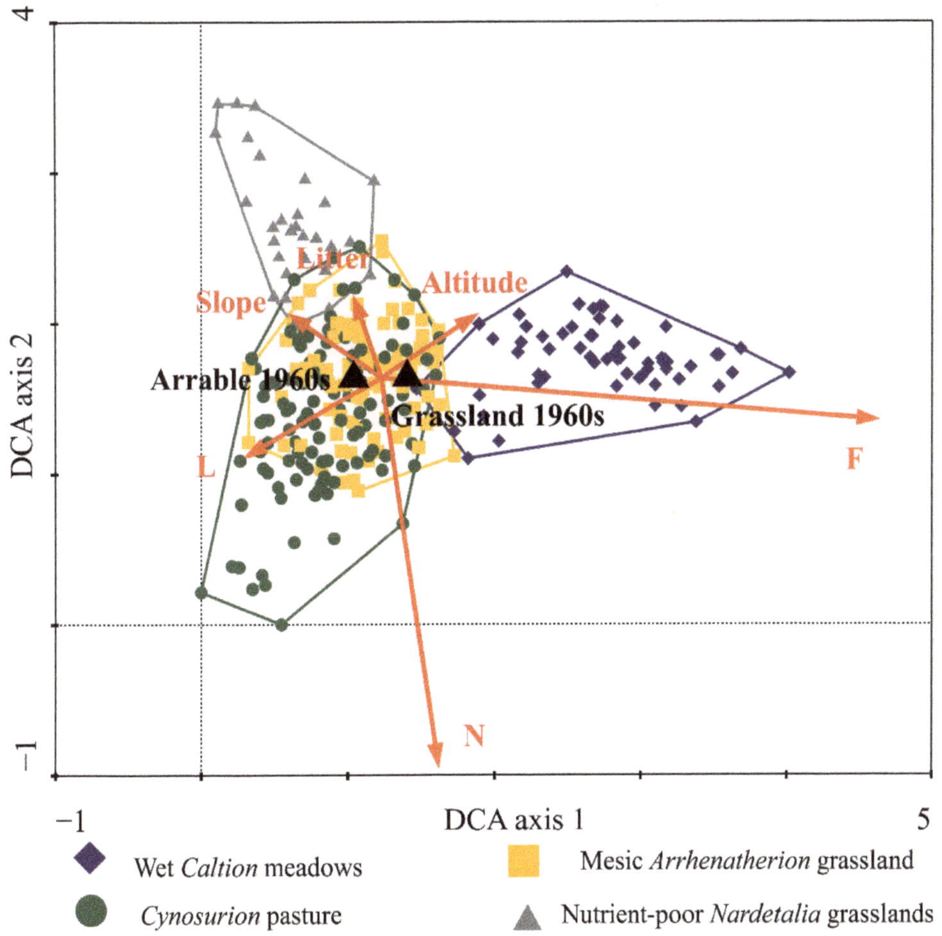

Figure 5. The Detrended Correspondence Analysis (DCA) ordination diagram of 294 plots categorized by grassland type with environmental vectors as overlay. Eigenvalues were 0.506 and 0.276 for the first and second axis, respectively. Cumulative percentage variance of species data for the first and the second axis were 4.9% and 7.5%, respectively. Ellenberg Indicator Values: F-moisture, L-light, N-nutrients.

Table 2. Diversity of grassland types in relation to land use in the past. Land use in the 1960s: G-grassland; A-arable fields.

Grassland Type	Land Use in the 1960s	No. of Relevés	Mean Species Richness (SD)	Mean Shannon Index (SD)
Wet *Caltion* meadows	G	47	27.5 (6.8)	2.58 (0.39)
	A	14	25.4 (8.0)	2.39 (0.43)
Mesic *Arrhenatherion* grassland	G	38	28.4 (7.9)	2.70 (0.40)
	A	53	30.0 (5.9)	2.80 (0.30)
Cynosurion pasture	G	55	22.7 (5.5)	2.46 (0.31)
	A	57	23.7 (6.2)	2.41 (0.37)
Nutrient-poor *Nardetalia* grassland	G	16	26.9 (7.5)	2.59 (0.34)
	A	14	27.4 (7.2)	2.55 (0.42)

The total number of species (gamma diversity) in grasslands situated in grazing sites ranged from 49 to 104, but only three sites were very poor in species (below 60). The average number of species per phytosociological relevé (alpha diversity) was 26.3, ranging from 17.6 to 34.3.

3.3. Conservation Status

The conservation status assessment of grasslands (Figure 6) showed the worst status of nutrient-poor *Nardetalia* grasslands, which were also rare in grazing sites and occupied the smallest surface areas. The conservation status of the wet meadows was assessed as favorable in one third of grazing areas. Mesic grasslands were present in almost all grazing areas; in more than half of grazing sites, their conservation status was evaluated as inadequate, in nine areas as bad and in eight areas as favorable. *Cynosurion* pastures were recorded in all grazing areas and their status was favorable in almost half, and was assessed as bad only at five sites.

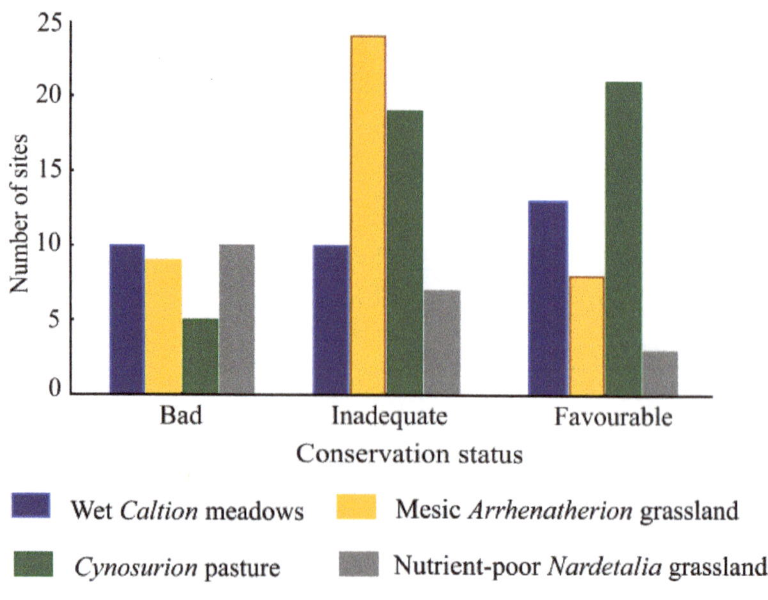

Figure 6. Conservation status of four grassland types evaluated on the basis of an assessment of their structure and function.

3.4. The Effect of Land Use on Species Diversity and Conservation Status of Grasslands

A QUAS index was associated with a greater total number of species in a grazing site and a higher number of species diagnostic of the *Cynosurion*, *Nardetalia*, and other grassland species, but with a lower number of species diagnostic of the *Calthion*. The number of species diagnostic of the *Cynosurion* and species diagnostic for the *Molinio-Arrhenatheretea* class and for the *Arrhentheretalia* order (*M-A+A*) was positively correlated with the share of arable fields in land use structure in the 1960s, while the number of species diagnostic of the *Calthion* was negatively correlated with this share. The impact of the current share of arable field area in land use structure was much greater and showed a positive correlation with the total number of species in a grazing site, the average number of species in a relevé, and the number of species diagnostic of the *Cynosurion*, grassland species and all other species, but had no effect on the number of species diagnostic of wet meadows (Table 3).

Table 3. Spearman rank-correlation coefficients between grazing site diversity metrics, habitat quality and grazing site characteristics. Abbreviations: *M-A+A*-species diagnostic for the *Molinio-Arrhenatheretea* class and for the *Arrhentheretalia* order.

	Soil Agriculture Suitability Index	Share of Arable Land Area 1960s	Share of Arable Land Area 2017
Diversity			
γ diversity	0.31 *	0.12	0.50 *
α diversity	0.28	0.06	0.34 *
Number of diagnostic species			
Calthion	−0.31 *	−0.31 *	−0.17
Arrhenatherion	0.19	0.22	0.27
Cynosurion	0.40 *	0.33 *	0.44 *
M-A+A	0.29 *	0.31 *	0.41 *
Nardetalia	0.32 *	−0.09	0.27
Others	0.30 *	0.06	0.48 *
Conservation status			
Wet *Caltion* meadows	−0.17	−0.30	−0.14
Mesic *Arrhenatherion* grassland	−0.13	−0.13	−0.15
Cynosurion pastures	0.14	0.02	0.03
Nutrient-poor *Nardetalia* grassland	−0.23	−0.26	−0.26

* significant $p < 0.05$.

4. Discussion

4.1. Species Composition and Diversity in Relation to Land Use Change

The results of the research indicate a process of increasing similarity in the species composition of grassland in areas where transhumant pastoralism is used. This applies in particular to communities of the alliances *Arrhenatherion* and *Cynosurion* because they can occur in the same environmental conditions, while the management type is a differentiating factor. Grazing caused the species composition of different communities to become similar. This is not the case for wet *Caltion* meadows and nutrient-poor *Nardetalia* grassland. The *Calthion* meadows occur in wet places and usually are only sporadically grazed. They are most often found in grazing sites with poor-quality soils at higher altitudes and with a slight proportion of arable fields in the past. Nutrient-poor *Nardetalia* grasslands in the Polish Carpathians are rare [42]. They most often cover small patches in habitat conditions differing from the surrounding grasslands. Thus, they can appear as enclaves in grazing areas among generally more fertile soils.

The highest biodiversity, as measured by the average number of all vascular species, pasture species and, generally, grassland species, was characteristic of grazing sites comprising a large proportion of arable fields. A significant proportion of arable fields in grazing sites in the past and their remains at present resulted in a high heterogeneity of habitats, related to the occurrence of such elements as field boundaries, clearance cairns, and dirt roads with roadsides. They are overgrown by plant species which increase species diversity in the landscape [32] and can easily migrate to grasslands. The heterogeneity of habitats and the related total number of species does not increase the abundance of grassland specialist species, since the latter occur in old grasslands [43,44]. Additionally, in our studies, in contrast to grassland generalist species (*M-A+A*), no correlation was found between the number of grassland specialist species and the proportion of arable fields in the grazing site.

The large number of pasture species in a grazing site was correlated with a higher soil quality index and with a higher share of arable fields both at present and in the past. It indicates that ex-arable fields create favorable conditions for pasture communities. Grazing enables the dispersion of species by endo- and epizoochory [45], while disturbances caused by trampling and browsing by animals increase the probability of successful recruitments [46]. Similar relationships were observed in the case of grassland generalist species. However, such use may be insufficient for grassland specialist species [47].

The relationship between present species composition and land use type in the 1960s is not strong. The land use data for that period allow only for the conclusion that grasslands on ex-arable fields do not exist for longer than 60 years. However, this period can be much shorter. Transformations occur gradually; thus, grasslands differ in age, which results in a variable species composition because the migration of propagules depends on time [47]. Öster et al. [48] indicated that over a 50-year period, only 50% of species occurring on neighboring semi-natural grasslands migrated to ex-arable fields. Additionally, investigations on dry grasslands demonstrated a significant role of former land use in terms of species distribution [49].

4.2. Conservation Status and Transhumant Pastoralism as a Protective Measure

Conservation status expressed by structure and function assessment index for all grassland types was not statistically significantly correlated with soil suitability for agriculture or present or past land-use type. Nevertheless, these are dynamic ecosystems in which land use-dependent vegetation changes are relatively fast. In another study [50], significant vegetation changes were found in abandoned mountain meadows only 6 years after the reinstatement of grazing, and after 9 years the proportion of species typical of floristically rich grasslands was increased. The interval of time analyzed by us (ca. 60 years) is probably too long to sustain a statistically significant impact of land use type at that time.

Evaluation of the restoration efficacy of semi-natural grasslands on former croplands should take into account the similarity of the species composition of created communities to the species composition of communities typical of local habitat conditions (referenced community), because it is a better indicator of success than species diversity (number of species) [51]. If no local reference communities have been preserved or the study areas are widely diverse, as in our case, it is necessary to strive for the achievement of conservation status, described as favorable for grassland communities, following the criteria formulated by State Environmental Monitoring [52–54].

The transformation of former arable fields into permanent grasslands is not currently a common practice in Europe, which is related to the extensification of land use, e.g., in protected areas, especially in the mountains [7], or the implementation of agri-environmental programs [55]. The restoration of multispecies communities on formerly intensively used arable fields is difficult and long lasting, and requires the application of diverse measures [56,57]. However, under favorable conditions, this process can be much faster [58]. In the Polish Carpathians, this process proceeds spontaneously through the migration of grassland species from neighboring areas to ex-arable fields. A majority of grazing sites are characterized by a mosaic of small patches of different land use types and by diverse vegetation; thus, an easily accessible source of propagules is available, which is crucial for grassland restoration [57,58]. High doses of fertilizer have never been used in the Polish Carpathians, while high-fertility soils are a significant obstacle to the restoration of species-rich grasslands [56]. A similar mechanism of preservation of high diversity was suggested by Janišová et al. [59] based on studies of semi-natural grasslands in Slovakia.

Traditional transhumance is practiced in the Carpathians, which involves livestock grazing in summer and the migration of herds within a large area covered by a mosaic of semi-natural grasslands, arable fields, ex-arable fields, trees and other small landscape features, allows species diversity to be preserved [59,60] because it supports natural processes, according to the metacommunity theory [61]. Extensive grazing can also produce negative effects from a nature conservation perspective if it is used in communities created by different management types. Grazing is considered to be more advantageous for species diversity than mowing [62]. However, many types of grasslands with high natural value, such as the mesic grasslands occurring in the study area, have developed and currently exist due to a proper mowing regime [63]. Plant diversity and conservation status largely depend on diversified land use, as revealed by the studies of Kun et al. [64] carried out in the Romanian Carpathians and Tölgyesi et al. [65] in Hungary. Hence, the preservation of the mosaic spatial structure of natural and semi-natural features and extensive farming practices in connection with diverse forms of human impact, such as grazing and mowing, is the optimal solution to preserve diversity at the levels of species, community, and landscape in mountain areas.

5. Conclusions

The transformation of extensively used arable fields into grasslands is relatively fast, but the species composition of communities formed in this way is dominated by less specialized species.

The biodiversity of analyzed grazing sites is dependent on the recent grassland area to arable land area ratio, rather than on the 60-year land use history. Nevertheless, detailed knowledge of past land use should be an integral part of analyses of the current state of vegetation and its dynamics. The spatial context (landscape connectivity, habitat fragmentation) is also worth considering.

Traditional transhumance can have a beneficial effect on the species diversity of grassland communities that have developed on former arable fields. Extensive open grazing over vast areas provides a real opportunity to maintain this type of ecosystem in good condition and should be subsidized by the state.

Author Contributions: Conceptualization, J.Z., J.K. and J.P.; methodology, J.Z., J.K. and J.P.; software, J.Z.; investigation, J.Z., J.K. and J.P.; writing—original draft preparation, J.Z., J.K. and J.P.; writing—review and editing, J.Z. and J.K.; visualization, J.K. and J.P.; project administration, J.P. All authors have read and agreed to the published version of the manuscript.

Funding: This manuscript is not funded by a specific project grant.

Data Availability Statement: The data presented in this study are available on request from the corresponding author.

Acknowledgments: The study was conducted in the framework of the statutory activity of the Institute of Nature Conservation, Polish Academy of Sciences and with a subsidy of the Ministry of Science and Higher Education for the University of Agriculture in Kraków in 2022. The results from field work carried out for the project "Maintenance of biological diversity of mountain meadows and pastures through pastoral management" were used. The project was realized by the Landscape Parks of the Malopolska Voivodship, co-financed by the European Regional Development Fund (ERDF) under the Regional Operational Program of the Malopolska Voivodship for the years 2014–2020.

Conflicts of Interest: The authors declare no conflict of interest.

References

1. Sollenberger, L.E.; Kohmann, M.M.; Dubeux, J.C.B.; Silveira, M.L. Grassland management affects delivery of regulating and supporting ecosystem services. *Crop Sci.* **2019**, *59*, 441–459. [CrossRef]
2. Bengtsson, J.; Bullock, J.M.; Egoh, B.; Everson, C.; Everson, T.; O'Connor, T.; O'Farrell, P.J.; Smith, H.G.; Lindborg, R. Grasslands—More important for ecosystem services than you might think. *Ecosphere* **2019**, *10*, e02582. [CrossRef]
3. Baura, B.; Cremene, C.; Groza, G.; Rakosy, L.; Schileyko, A.A.; Baura, A.; Stoll, P.; Erhardt, A. Effects of abandonment of subalpine hay meadows on plant and invertebrate diversity in Transylvania, Romania. *Biol. Conserv.* **2006**, *132*, 261–273. [CrossRef]
4. Leuschner, C.; Ellenberg, H. *Ecology of Central European Non-Forest Vegetation: Coastal to Alpine, Natural to Man-Made Habitats;* Vegetation Ecology of Central Europe Series; Springer International Publishing: Cham, Switzerland, 2017; Volume II.

5. Gilhaus, K.; Boch, S.; Fischer, M.; Hölzel, N.; Kleinebecker, T.; Prati, D.; Rupprecht, D.; Schmitt, B.; Klaus, V.H. Grassland management in Germany: Effects on plant diversity and vegetation composition. *Tuexenia* **2017**, *37*, 379–397. [CrossRef]

6. Bakker, M.; Veldkamp, A. Changing relationships between land use and environmental characteristics and their consequences for spatially explicit land-use change prediction. *J. Land Use Sci.* **2012**, *7*, 407–424. [CrossRef]

7. Batáry, P.; Dicks, L.V.; Kleijn, D.; Sutherland, W.J. The role of agri-environment schemes in conservation and environmental management. *Conserv. Biol.* **2015**, *29*, 1006–1016. [CrossRef]

8. Cousins, S.A.O.; Auffret, A.G.; Lindgren, J.; Tränk, L. Regional-scale land-cover change during the 20th century and its consequences for biodiversity. *Ambio* **2015**, *44*, 17–27. [CrossRef] [PubMed]

9. Tasser, E.; Tappeiner, U. Impact of land use changes on mountain vegetation. *Appl. Veg. Sci.* **2002**, *5*, 173–184. [CrossRef]

10. Wesche, K.; Krause, B.; Culmsee, H.; Leuschner, C. Fifty years of change in Central European grassland vegetation: Large losses in species richness and animal-pollinated plants. *Biol. Conserv.* **2012**, *150*, 76–85. [CrossRef]

11. Zarzycki, J.; Bedla, D. The influence of past land-use and environmental factors on grassland species diversity. *Appl. Ecol. Environ. Res.* **2017**, *15*, 267–278. [CrossRef]

12. Chabuz, W.; Kulik, M.; Sawicka-Zugaj, W.; Żółkiewski, P.; Warda, M.; Pluta, M.; Lipiec, A.; Bochniak, A.; Zdulski, J. Impact of the type of use of permanent grasslands areas in mountainous regions on the floristic diversity of habitats and animal welfare. *Glob. Ecol. Conserv.* **2019**, *19*, e00629. [CrossRef]

13. Fischer, M.; Wipf, S. Effect of low-intensity grazing on the species-rich vegetation of traditionally mown subalpine meadows. *Biol. Conserv.* **2002**, *104*, 1–11. [CrossRef]

14. Kolecka, N.; Kozak, J.; Kaim, D.; Dobosz, M.; Ostafin, K.; Ostapowicz, K.; Wężyk, P.; Price, B. Understanding farmland abandonment in the Polish Carpathians. *Appl. Geogr.* **2017**, *88*, 62–72. [CrossRef]

15. Twardy, S. Karpackie użytki rolne jako obszary o niekorzystnych warunkach gospodarowania (ONW). *Woda Śr. Obsz. Wiej.* **2008**, *24*, 181–192.

16. Gracheva, R.; Belonovskaya, E.; Vinogradova, V. Mountain grassland ecosystems on abandoned agricultural terraces (Russia, North Caucasus). *Hacquetia* **2018**, *17*, 61–71. [CrossRef]

17. Pazúr, R.; Lieskovský, J.; Feranec, J.; Oťaheľ, J. Spatial determinants of abandonment of large-scale arable lands and managed grasslands in Slovakia during the periods of post-socialist transition and European Union accession. *Appl. Geogr.* **2014**, *54*, 118–128. [CrossRef]

18. Ferrara, A.; Biró, M.; Malatesta, L.; Molnár, Z.; Mugnoz, S.; Tardella, F.M.; Catorci, A. Land-use modifications and ecological implications over the past 160 years in the central Apennine mountains. *Landsc. Res.* **2021**, *46*, 932–944. [CrossRef]

19. Cousins, S.A.O.; Aggemyr, E. The influence of field shape, area and surrounding landscape on plant species richness in grazed ex-fields. *Biol. Conserv.* **2008**, *141*, 126–135. [CrossRef]

20. Waesch, G.; Becker, T. Plant diversity differs between young and old mesic meadows in a central European low mountain region. *Agric. Ecosyst. Environ.* **2009**, *129*, 457–464. [CrossRef]

21. Gustavsson, E.; Lennartsson, T.; Emanuelsson, M. Land use more than 200 years ago explains current grassland plant diversity in a Swedish agricultural landscape. *Biol. Conserv.* **2007**, *138*, 47–59. [CrossRef]

22. Kuhn, T.; Domokos, P.; Kiss, R.; Ruprecht, E. Grassland management and land use history shape species composition and diversity in Transylvanian semi-natural grasslands. *Appl. Veg. Sci.* **2021**, *24*, e12585. [CrossRef]

23. Reitalu, T.; Johansson, L.J.; Sykes, M.T.; Hall, K.; Prentice, H.C. History matters: Village distances, grazing and grassland species diversity. *J. Appl. Ecol.* **2010**, *47*, 1216–1224. [CrossRef]

24. Berezowski, S. Problemy geograficzne pasterstwa wędrownego [Geographical Problems of Nomadic Pastoralism]. In *Pasterstwo Tatr i Podhala*; Antoniewicz, W., Ed.; Zakład Narodowy Imienia Ossolińskich: Wrocław, Poland, 1959; pp. 77–146.

25. Huband, S.; Mccracken, D.I.; Mertens, A. Long and short-distance transhumant pastoralism in Romania: Past and present drivers of change. *Pastor. Res. Policy Pract.* **2010**, *1*, 55–71. [CrossRef]

26. Warchalska-Troll, A.; Troll, M. Summer Livestock Farming at the Crossroads in the Ukrainian Carpathians. *Mt. Res. Dev.* **2014**, *34*, 344–355. [CrossRef]

27. Liechti, K.; Biber, J.P. Pastoralism in Europe: Characteristics and challenges of highland-lowland transhumance. *OIE Rev. Sci. Tech.* **2016**, *35*, 561–575. [CrossRef]

28. Główny Urząd Statystyczny. *Statistical Yearbook of Agriculture 2018*; Główny Urząd Statystyczny: Warsaw, Poland, 2019.

29. Sendyka, P.; Makovicky, N. Transhumant pastoralism in Poland: Contemporary challenges. *Pastoralism* **2018**, *8*, 5. [CrossRef]

30. De Blois, S.; Domon, G.; Bouchard, A. Landscape issues in plant ecology. *Ecography* **2002**, *25*, 244–256. [CrossRef]

31. Lindborg, R.; Eriksson, O. Historical Landscape Connectivity Affects Present Plant Species Diversity. *Ecol. Soc. Am.* **2004**, *85*, 1840–1845. [CrossRef]

32. Cousins, S.A.O. Plant species richness in midfield islets and road verges—The effect of landscape fragmentation. *Biol. Conserv.* **2006**, *127*, 500–509. [CrossRef]

33. Braun-Blanquet, J. *Pflanzensoziologie. Grundzüge der Vegetationskunde*, 3rd ed.; Springer: Vienna, Austria, 1964.

34. Chief Inspectorate for Environmental Protection. 2017. Available online: http://siedliska.gios.gov.pl/pl/publikacje/przewodniki-metodyczne/methodological-guides (accessed on 28 June 2021).

35. European Commission. *Interpretation Manual of European Union Habitats*; European Commission, DG Environment: Brussels, Belgium, 2007.

36. Mirek, Z.; Piękoś-Mirek, H.; Zając, A.; Zając, M. *Flowering Plants and Pteridophytes of Poland. A Checklist*; W. Szafer Institute of Botany, Polish Academy of Sciences: Kraków, Poland, 2002.

37. Knollová, I.; Chytrý, M.; Tichý, L.; Hájek, O. Stratified resampling of phytosociological databases: Some strategies for obtaining more representative data sets for classification studies. *J. Veg. Sci.* **2005**, *16*, 479–486. [CrossRef]

38. Tichý, L. JUICE, software for vegetation classification. *J. Veg. Sci.* **2002**, *13*, 451–453. [CrossRef]

39. Witek, T. *Waloryzacja Rolniczej Przestrzeni Produkcyjnej Polski Według Gmin*; IUNG: Puławy, Poland, 1981.

40. Matuszkiewicz, W. *Klucz do Oznaczania Zbiorowisk Roślinnych Polski*; Wydawnictwo Naukowe PWN: Warsaw, Poland, 2011.

41. ter Braak, C.J.F.; Smilauer, P. *Canoco Reference Manual and User's Guide: Software for Ordination Version 5.1*; Microcomputer Power: Ithaca, NY, USA, 2018.

42. Korzeniak, J. Mountain *Nardus stricta* grasslands as a relic of past farming—The effects of grazing abandonment in relation to elevation and spatial scale. *Folia Geobot.* **2016**, *51*, 93–113. [CrossRef]

43. Cousins, S.A.O.; Eriksson, O. The influence of management history and habitat on plant species richness in a rural hemiboreal landscape, Sweden. *Landsc. Ecol.* **2002**, *17*, 517–529. [CrossRef]

44. Winsa, M.; Bommarco, R.; Lindborg, R.; Marini, L.; Öckinger, E. Recovery of plant diversity in restored semi-natural pastures depends on adjacent land use. *Appl. Veg. Sci.* **2015**, *18*, 413–422. [CrossRef]

45. Baltzinger, C.; Karimi, S.; Shukla, U. Plants on the move: Hitch-hiking with ungulates distributes diaspores across landscapes. *Front. Ecol. Evol.* **2019**, *7*, 1–19. [CrossRef]

46. Jakobsson, A.; Eriksson, O. A Comparative Study of Seed Number, Seed Size, Seedling Size and Recruitment in Grassland Plants. *Oikos* **2000**, *88*, 494–502. [CrossRef]

47. Waldén, E.; Öckinger, E.; Winsa, M.; Lindborg, R. Effects of landscape composition, species pool and time on grassland specialists in restored semi-natural grasslands. *Biol. Conserv.* **2017**, *214*, 176–183. [CrossRef]

48. Öster, M.; Ask, K.; Römermann, C.; Tackenberg, O.; Eriksson, O. Plant colonization of ex-arable fields from adjacent species-rich grasslands: The importance of dispersal vs. recruitment ability. *Agric. Ecosyst. Environ.* **2009**, *130*, 93–99. [CrossRef]

49. Chýlová, T.; Münzbergová, Z. Past land use co-determines the present distribution of dry grassland plant species. *Preslia* **2008**, *80*, 183–198.

50. Krahulec, F.; Skálová, H.; Herben, T.; Hadincov, V.; Wildová, R.; Pecháčková, S. Vegetation changes following sheep grazing in abandoned mountain meadows. *Appl. Veg. Sci.* **2001**, *4*, 97–102. [CrossRef]

51. Waldén, E.; Lindborg, R. Long term positive effect of grassland restoration on plant diversity—Success or not? *PLoS ONE* **2016**, *11*, 1–16. [CrossRef] [PubMed]

52. Mróz, W. (Ed.) *Monitoring of Natural Habitats. Methodological Guide*; Biblioteka Monitoringu Środowiska, GIOŚ: Warsaw, Poland, 2010; Volume 1.

53. Mróz, W. (Ed.) *Monitoring of Natural Habitats. Methodological Guide*; Biblioteka Monitoringu Środowiska, GIOŚ: Warsaw, Poland, 2012; Volume 2.

54. Mróz, W. (Ed.) *Monitoring of Natural Habitats. Methodological Guide*; Biblioteka Monitoringu Środowiska, GIOŚ: Warsaw, Poland, 2015; Volume 3.

55. Stoate, C.; Báldi, A.; Beja, P.; Boatman, N.D.; Herzon, I.; van Doorn, A.; de Snoogh, G.R.; Rakosy, L.; Ramwell, C. Ecological impacts of early 21st century agricultural change in Europe—A review. *J. Environ. Manag.* **2009**, *91*, 22–46. [CrossRef] [PubMed]

56. Kiehl, K.; Kirmer, A.; Donath, T.W.; Rasran, L.; Hölzel, N. Species introduction in restoration projects—Evaluation of different techniques for the establishment of semi-natural grasslands in Central and Northwestern Europe. *Basic Appl. Ecol.* **2010**, *11*, 285–299. [CrossRef]

57. Török, P.; Vida, E.; Deák, B.; Lengyel, S.; Tóthmérész, B. Grassland restoration on former croplands in Europe: An assessment of applicability of techniques and costs. *Biodivers. Conserv.* **2011**, *20*, 2311–2332. [CrossRef]

58. Ruprecht, E. Successfully recovered grassland: A promising example from Romanian old-fields. *Restor. Ecol.* **2006**, *14*, 473–480. [CrossRef]

59. Janišová, M.; Michalcová, D.; Bacaro, G.; Ghisla, A. Landscape effects on diversity of semi-natural grasslands. *Agric. Ecosyst. Environ.* **2014**, *182*, 47–58. [CrossRef]

60. Kumm, K.I. Does re-creation of extensive pasture-forest mosaics provide an economically sustainable way of nature conservation in Sweden's forest dominated regions? *J. Nat. Conserv.* **2004**, *12*, 213–218. [CrossRef]

61. Leibold, M.A.; Holyoak, M.; Mouquet, N.; Amarasekare, P.; Chase, J.M.; Hoopes, M.F.; Holt, R.D.; Shurin, J.B.; Law, R.; Tilman, D.; et al. The metacommunity concept: A framework for multi-scale community ecology. *Ecol. Lett.* **2004**, *7*, 601–613. [CrossRef]

62. Tälle, M.; Deák, B.; Poschlod, P.; Valkó, O.; Westerberg, L.; Milberg, P. Grazing vs. mowing: A meta-analysis of biodiversity benefits for grassland management. *Agric. Ecosyst. Environ.* **2016**, *222*, 200–212. [CrossRef]

63. Zarzycki, J.; Korzeniak, J. Łąki w polskich Karpatach—Stan aktualny, zmiany i możliwości ich zachowania. *Rocz. Bieszcz.* **2013**, *21*, 18–34.
64. Kun, R.; Bartha, S.; Malatinszky, A.; Molnár, Z.; Lengyel, A.; Babai, D. "Everyone does it a bit differently!": Evidence for a positive relationship between micro-scale land-use diversity and plant diversity in hay meadows. *Agric. Ecosyst. Environ.* **2019**, *283*, 106556. [CrossRef]
65. Tölgyesi, C.; Török, P.; Kun, R.; Csathó, A.I.; Bátori, Z.; Erdős, L.; Vadász, C. Recovery of species richness lags behind functional recovery in restored grasslands. *Land Degrad. Dev.* **2019**, *30*, 1083–1094. [CrossRef]

 *land*

Article

Influence of Anthropogenic Load in River Basins on River Water Status: A Case Study in Lithuania

Laima Česonienė [1,*], Daiva Šileikienė [1] and Midona Dapkienė [2]

[1] Department of Environment and Ecology, Faculty of Forest Science and Ecology, Vytautas Magnus University, Agriculture Academy, Studentų Str. 11, LT–53361 Akademija, Lithuania; daiva.sileikiene@vdu.lt

[2] Department of Water Engineering, Faculty of Engineering, Vytautas Magnus University, Agriculture Academy, Universiteto Str., 10, LT–53361 Akademija, Lithuania; midona.dapkiene@vdu.lt

* Correspondence: laima.cesoniene1@vdu.lt; Tel.: +370-37-752224

Abstract: Twenty-four rivers in different parts of Lithuania were selected for the study. The aim of the research was to evaluate the impact of anthropogenic load on the ecological status of rivers. Anthropogenic loads were assessed according to the pollution sources in individual river catchment basins. The total nitrogen (TN) values did not correspond to the "good" and "very good" ecological status classes in 51% of the tested water bodies; 19% had a "bad" to "moderate" BOD_7, 50% had "bad" to "moderate" NH_4-N, 37% had "bad" to "moderate" NO_3-N, and 4% had "bad" to "moderate" PO_4-P. The total phosphorus (TP) values did not correspond to the "good" and "very good" ecological status classes in 4% of the tested water bodies. The largest amounts of pollution in river basins were generated from the following sources: transit pollution, with 87,599 t/year of total nitrogen and 5020 t/year of total phosphorus; agricultural pollution, with 56,031 t/year of total nitrogen and 2474 t/year of total phosphorus. The highest total nitrogen load in river basins per year, on average, was from transit pollution, accounting for 53.89%, and agricultural pollution, accounting for 34.47%. The highest total phosphorus load was also from transit pollution, totaling 58.78%, and agricultural pollution, totaling 28.97%. Multiple regression analysis showed the agricultural activity had the biggest negative influence on the ecological status of rivers according to all studied indicators.

Keywords: pollution; ecological status indicators; water quality

Citation: Česonienė, L.; Šileikienė, D.; Dapkienė, M. Influence of Anthropogenic Load in River Basins on River Water Status: A Case Study in Lithuania. *Land* **2021**, *10*, 1312. https://doi.org/10.3390/land10121312

Academic Editors: Wiktor Halecki, Dawid Bedla, Marek Ryczek and Artur Radecki-Pawlik

Received: 25 October 2021
Accepted: 26 November 2021
Published: 28 November 2021

Publisher's Note: MDPI stays neutral with regard to jurisdictional claims in published maps and institutional affiliations.

Copyright: © 2021 by the authors. Licensee MDPI, Basel, Switzerland. This article is an open access article distributed under the terms and conditions of the Creative Commons Attribution (CC BY) license (https://creativecommons.org/licenses/by/4.0/).

1. Introduction

Lithuania is committed to achieving the objectives of the EU Water Framework Directive by 2027 and to achieving good water status in inland waters. There are approximately 30 thousand rivers and creeks in Lithuania, with a longer than 200 m, reaching an overall sum of 63,700 km. Although the ecological condition of Lithuanian rivers has been mostly improving over the last few years, it has been determined that only 49% of them correspond to a good ecological state [1], in the period of 2010–2013.

Human activities change the hydromorphological, physicochemical, and biological parameters of surface water bodies, affecting their biodiversity and ecological functioning [2–9]. Pollution caused by anthropogenic activity sufficiently worsens the state of water ecosystems, resulting in water quality degradation, reducing the potential for water use in various fields and endangering people health [10–14].

Sources of pollution are divided as follows: background pollution (forests), diffuse (non-point) pollution from agricultural lands, surface sewage that is not treated in wastewater treatment plants (WWTPs), and concentrated (point) pollution caused by households, urban, municipal, industrial wastewater (wastewater treatment plants), and others. Human activities affect the condition of water bodies differently in rural areas (agricultural activities, livestock) and urban areas (industrial, municipal, domestic wastewater discharges). Changes of land use also have a negative effect on water condition of rivers [15–21]. Landscape changes caused by anthropogenic activities and land cover make a significant influence on the state of surface waters [22–24], are closely related to water chemical parameters [25], the diversity of fish and macroinvertebrate species [24,26], and sediment metal concentrations [27]. Fragmented urban land use, with many impermeable surfaces, tends to increase river flows and adversely affects water quality [28–30].

Many research have been conducted to evaluate the ecological status of surface waters [31–36]. The multiple anthropogenic pressures (pollution, hydrological, and hydromorphological alterations) on the ecological status of the European rivers were assessed. In one third of the territory of the European Union, the ecological status of rivers has been found to be good, which is linked to the existence of natural areas, and urbanization is leading to poorer ecological status [37].

A concentrated source of pollution is wastewater from factories and households, and it is insufficiently treated and controlled [38–40]. Joshua et al. (2017) have pointed out that substandard practices of wastewater treatment are utilized in developing countries. In these countries, the major causes for this include the insufficient number of wastewater treatment plants, overloading and ineffective operation of existing WWTPs, and others [41].

Diffuse pollution is one of the main problems in irrigated areas. Intensive irrigation and fertilization of arable land and pastures have the greatest negative impact [42,43].

The sewage from residents and their house-holds' lack of connection to the wastewater collection networks also forms diffuse pollution that enters rivers [44]. Diffuse pollution covers large areas, and all such areas are polluted quite equally [45]. Very important diffuse pollution source affecting the condition of rivers is inadequate farming [46–49]. The riverbanks are a suitable place for agriculture, as they are particularly productive. Diffuse pollution is caused not only by farming on riverbanks, but also by agricultural activities in all the river basin. The use of chemical fertilizers and pesticides is an essential part of diffuse pollution entering surface water bodies [50]. Both organic and mineral fertilizers are used to fertilize crops; however, due to erosion, soil leaching, and runoff, they enter surface water bodies and contaminate them with nutrients [51]. Agriculture was the cause of 48% of the deterioration in water quality of surface water bodies of USA [52]. Approximately 30–35% of nitrogen and 10–15% of phosphorus, which pollute surface waters, have been found to come from agricultural activities [53].

Generally, Lithuanian surface water bodies are impacted by both diffuse and concentrated source pollution. In Lithuania, the effect of pollution on the condition of Venta and Mūša-Lielupė river basins was evaluated [38,54,55]. According to the biochemical oxygen demand (BOD_7), human activity accounted for 56% of the influence of pollution on water quality, and 90% of the annual borne total nitrogen (TN) and 78% of the total phosphorus (TP) content in the Merkys River [56]. Of the bodies of water, 46% did not achieve a "good" status in terms of nitrate nitrogen in 2017 [57]. Export coefficients and the retention of biogenic nutrients in Lithuanian river basins were assessed using the MESAW statistical model. The export coefficients of TN and TP showed much higher values from arable land in comparison to forest area, pastures, and meadows from the studied Merkys, Mūša, Žeimena, and Nevėžis river basins, and retained from 67% to 78% of the total nitrogen and from 24% to 63% of the total phosphorus [58].

We performed a study of the effectiveness of diffuse pollution abatement measures in reducing the nutrient pollution of surface water bodies in Lithuania in the context of climate change. The SWAT model was used for this purpose. The results show that climate change is a significant factor in changing the effectiveness of measures to reduce diffuse pollution. The most effective measures to reduce nutrients inputs to water bodies were identified, including pasture/meadow expansion, stubble abandonment for winter, and catch crop cultivation; arable farming was the least efficient method [59].

The aim of this research is to evaluate the impact of anthropogenic load on indicators of the ecological status of rivers.

2. Materials and Methods

2.1. Study Area

To determine the risk of water bodies that do not comply with the water quality standards, the physicochemical quality indicators at 94 locations of 24 rivers were studied. Water samples were taken between January and March, April and June, July and September, and October and December in 2014–2020. The investigated river's water sampling areas and their hydrological data are presented in Figure 1 and Table 1.

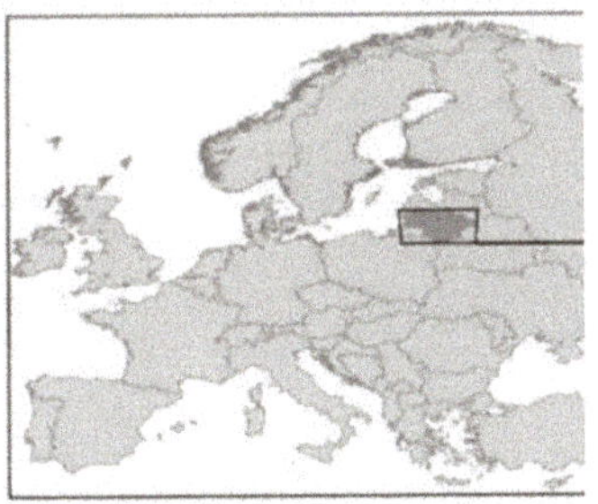

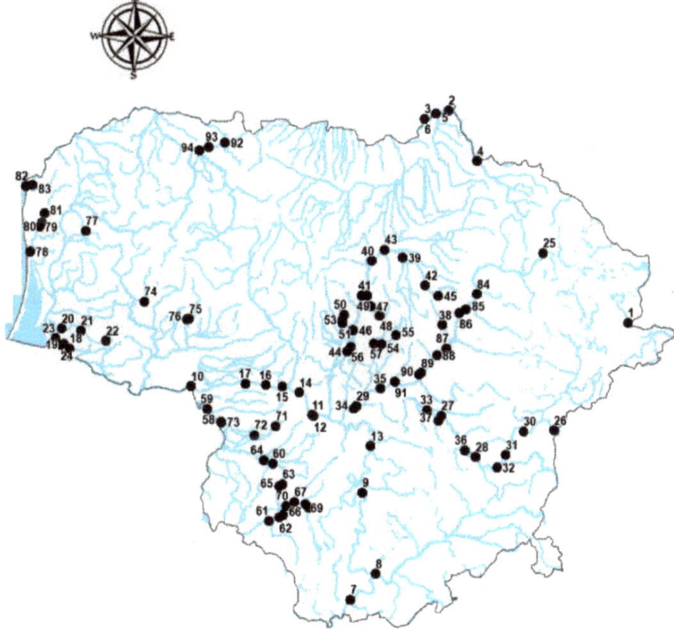

Figure 1. River's water sampling areas.

Table 1. Hydrological data of rivers.

River	Length, km	Length in Lithuania, km	Basin Area, km^2	Basin Area in Lithuania, km^2	Average Flow Rate, m^3/s	Number of Sampling Points	Number on the Map (Figure 1)
Dysna	176	77	8193	726	3.59	1	1
Nemunėlis	191	151	4048	3770	95	5	2–6
Nemunas	937.4	359	98,200	46,600	540	11	7–17
Leitė	26.2	26.2	143	143	1.5	2	18–19
Šyša	61	61	410	410	1.88	3	20–22
Skirvytė	9					2	23–24
Šventoji	246	246	6888.8	6800.7	55.1	11	25; 82–91
Neris	510	237.8	24,942.3	1392	181	10	26–35
Bražuolė	22.7	22.7	109.4	109.4	0.71	1	36
Žiežmara	24	24	65	65	0.49	1	37
Mušia	29	29	227.3	227.3	1.69	1	38
Nevėžis	209	209	6140	6140	33.2	8	39–46
Linkava	36.8	36.8	163.4	163.4	0.82	3	47–49
Kruostas	28.9	28.9	99.7	99.7	0.5	4	50–53
Obelis	53.3	53.3	673.8	673.8	2.7	4	54–57
Šešupė	297.6	297.6	6104.8	4899	34.2	9	58–66
Dovinė	65	65	588.7	588.7	3.4	4	67–70
Nova	69	69	403	403	1.24	3	71–73
Lokysta	46.3	46.3	173	173	2.12	1	74
Ančia	66.4	66.4	278.6	278.6	2.82	1	75
Agluona	22	22	76	76	0.98	1	76
Alantas	43	43	146	146	181	1	77
Akmena—Danė	62.5	62.5	595	595	6.9	4	78–81
Dabikinė	37.2		387.6		2.39	3	92–94

2.2. Water Quality Assessment Standards

The ecological statuses of water bodies at risk were assessed in accordance with the Procedure for Rating the Ecological Status of Surface Water Basin [60]. Research on the physical and chemical quality of elemental indicators have been identified in a laboratory at Vytautas Magnus University. Water samples were collected according to the EN ISO requirement standards: LST EN ISO 5667-14:2016—Water Quality—Sampling—Part 14. The total nitrogen (TN) was tested according to the method of LST EN 13342—2002. The total nitrogen (TN) following determination of nitrogen—determination of bound nitrogen (TNb), following oxidation to nitrogen oxides LST EN 12260:2004. Total phosphorus (TP) analyses were assessed according to LST EN ISO 6878:2004; BOD$_7$—according to ISO 5815-1:2003 ammonium nitrogen (NH$_4^+$-N) LST ISO 7150-1:1998, LST EN ISO 13395:2000 nitrate nitrogen (NO$_3$-N), and LST EN ISO 6878:2004 phosphate phosphorus (PO$_4$-P).

2.3. Presentation of Pollution Sources

The assessment of contamination sources considered the nature of land use, the nature of cities and settlements, the location of potential sources of concentrated source pollution, the nature and intensity of economic activities in the basin and their potential impact on water bodies, recreational activities, and other economic activities that may not be in good condition according to condition requirements, and so forth.

Diffuse agricultural pollution, consisting of manure and mineral fertilizer loads resulting from agricultural activities and from the load on the population whose households are not connected to sewage collection systems.

The main sources of concentrated pollution are wastewater from cities, settlements, industrial enterprises and rain and surface water, wastewater from urban areas.

High potential for concentrated pollution to enter water bodies directly or through river tributaries.

For the quantification of pollution indicators, the following factors have been assessed:

- Domestic and industrial wastewater disposal facilities in the study areas, the extent of their pollution loads, the impact on the status of the water body and the average TN and TP value of wastewater in the period 2015–2020. Data from the Environmental Protection Agency (EPA) on wastewater dischargers, identified pollutant concentrations, and annual wastewater volumes were estimated by dividing their statistical values by water body feeding basins;
- The number of people connected to the sewage collection systems and sewage management (i.e., central, individual, or no management (statistics)). The contamination loads in the environment released by the residents whose wastewater was not collected were assessed according to the HELCOM recommendations, which specify that one resident generates 25.6 kg of waste according to the BOD_7, 4.4 kg of TN, and 0.9 kg of TP;
- To determine the nutrient loads from residential and commercial areas, data from the SWAT (small watershed to river basin-scale model used to simulate the quality and quantity of surface and ground water) model were used to calculate and evaluate pollution loads. SWAT model is a basin-scale continuous-time model that operates on a continuous basis and assesses the impact of management practices on water, sediment, and agrochemicals in non-monitored basins [61]. SWAT is widely used in assessing soil erosion prevention and control, diffuse source pollution control and regional management in watersheds;
- To assess the impact of the transformation of biogenic nutrients in soil and water body pollution, a SWAT model was used to calculate the average of total nitrogen (TN) and total phosphorus (TP) leaching.

2.4. Statistical Analyses

To statistically assess the significant impacts on quality factors related to the ecological status of water bodies, the impacts of anthropogenic load indicators TP and TN from municipal wastewater, surface wastewater, households not connected to sewage networks, agricultural land, background, and transit pollution (t/year); agricultural land, forests, wetlands, meadows, arable, infertile land, and green land (ha) on water quality indicators (Y) for the water in rivers were determined. A multiple linear regression model was applied:

$$Y = a + b_1 x_1 + b_2 x_2 + \ldots + b_k x_k. \tag{1}$$

The coefficient b_j shows how much the value of Y increases (or decreases) by one unit, as x_j increases when the remaining x_k are fixed. t is Student's criterion, according to which we determined whether the b_j coefficients differed statistically significantly from zero, and according to this, we decided whether the predicted values depended upon x_j. The standardized coefficient beta was used to determine the relative influence of independent variables on the predicted Y. In absolute terms, a higher beta coefficient indicates greater dependence of Y on x_j.

The regression model is appropriate due to the following:

- The Levene test was applied as an endogeneity test; the R code was applied to generate the analyses in this area, $R^2 \geq 0.20$;
- ANOVA was performed with a *p*-value of <0.05;
- *t*-tests were performed, showing significance at $p < 0.05$;
- All SWFs (Dispersion reduction factor) were ≤ 4 (no diversity problems);
- All Cook measure values were ≤ 1.

3. Results

3.1. Ecological Status Classes of the Stretches of Rivers According to the Physicochemical Values of Elemental Indicators

Studies on the physicochemical quality of element indicators were performed for NO_3-N (mg/L), NH_4-N (mg/L), TN (mg/L), PO_4-P (mg/L), TP (mg/L), and the BOD_7 (mg/L). The results are shown in Figure 2.

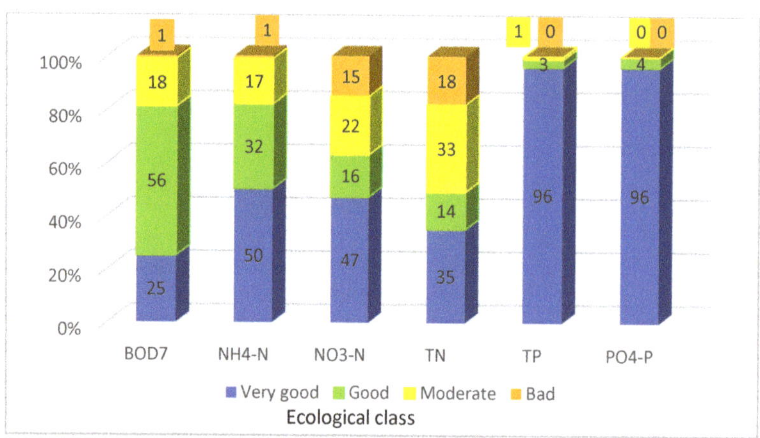

Figure 2. Ecological status classes of rivers according to the values of indicators of the physicochemical quality of elements%.

The results presented in Figure 2 show that, according to the TN, 51% of the studied rivers did not meet the requirements of the "good" ecological class; 19% of the rivers had a "bad" to "moderate" BOD_7, 50% had "bad" to "moderate" NH_4-N, 37% had "bad" to "moderate" NO_3-N, 4% had "bad" to "moderate" TP, and 4% had "bad" to "moderate" PO_4-P.

3.2. Assessment of Nutrient Loads in River Basins

Nutrient loads in the river basins were calculated by collected the SWAT model data. Calculations were performed in tons per year for the inflows into the rivers for the total nitrogen and total phosphorus. The TN and TP loads in river basins (t/year) are presented in Figure 3.

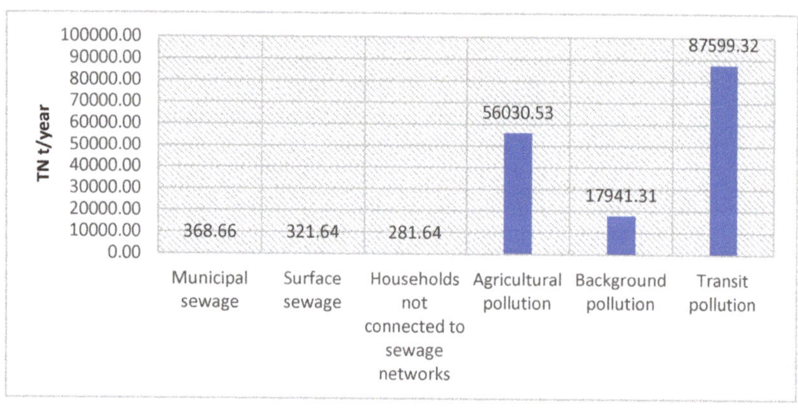

Figure 3. *Cont.*

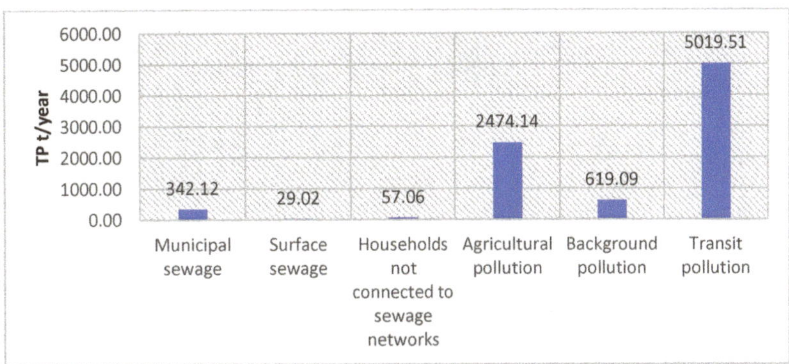

Figure 3. Amounts of total nitrogen and total phosphorus load in river basins (t/year).

The river basins get the largest amounts of pollution from transit loads located above the research locations, with the total nitrogen equaling 87,599.32 t/year and total phosphorus amounting to 5019.51 t/year. From agricultural activities, the total nitrogen reached 56,030.53 t/year and the total phosphorus was 2474.14 t/year. The amounts from background pollution (urban areas and forests) were 17,941.31 t/year of total nitrogen and 619.09 t/year of phosphorus. The amounts from municipal sewage were 368.66 t/year of total nitrogen and 342.12 t/year of phosphorus. The amounts from surface sewage were 321.64 t/year of total nitrogen and 29.02 t/year of phosphorus. Residents whose sewage was not discharged into sewage treatment systems generated 281.64 t/year of total nitrogen and 57.06 t/year of phosphorus.

Figure 4 shows the percentage distribution of total nitrogen and total phosphorus loads in the studied river basins.

The highest annual total nitrogen load for river basins per year, on average, came from transit pollution, accounting for 53.89%. A total of 34.47% came from agricultural pollution, 11.04% came from background pollution (urban areas and forests), 0.17% came from pollution from residents who were not connected to sewage systems, 0.20% came from surface sewage, and 0.23% came from municipal wastewater.

The highest annual load of total phosphorus in river basins was from transit pollution, accounting for 58.78%. A total of 28.97% came from agricultural pollution, 7.25% came from background pollution (urban areas and forests), 0.67% came from pollution from inhabitants who were not connected to sewage systems, 0.34% came from surface sewage, and 4.01% came from municipal wastewater.

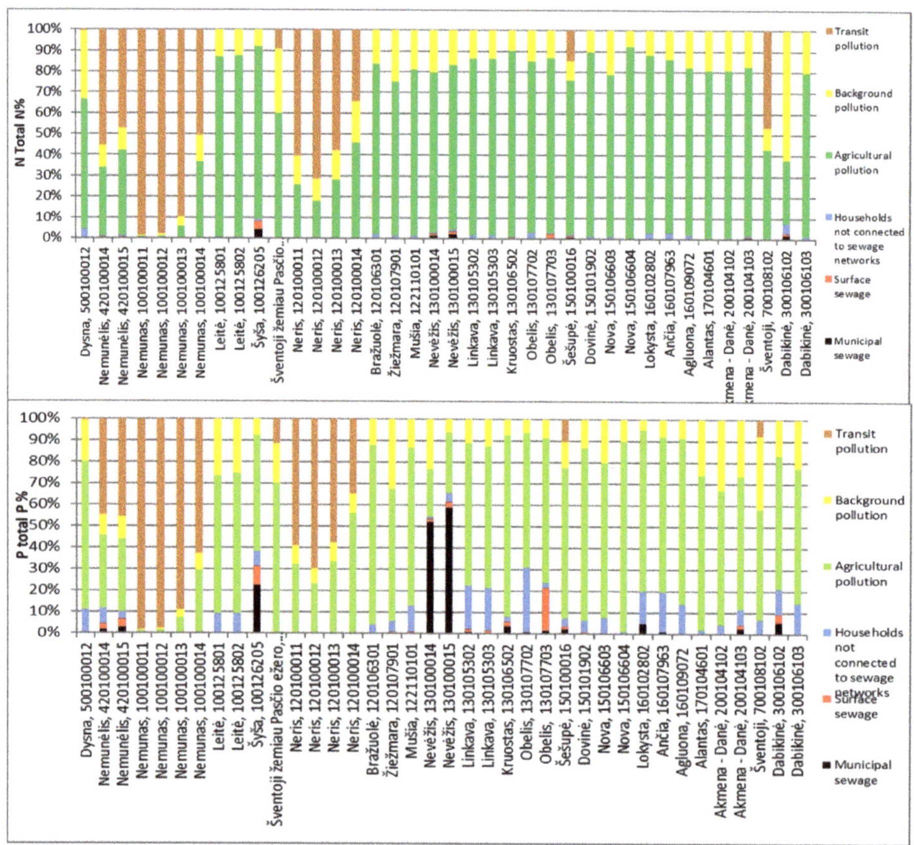

Figure 4. Percentage distribution of total nitrogen and total phosphorus loads in studied river basins (names of rivers and water body code).

3.3. Influence of Anthropogenic Loading on Total Nitrogen, Ammonium Nitrogen, Nitrate Nitrogen, Total Phosphorus, and Phosphate Phosphorus

The influence of anthropogenic loading on total nitrogen concentration (TN is dependent variable Y) was calculated by multiple regression analysis and results are presented in Table 2.

Multiple regression analysis of the influence of anthropogenic loads on the total nitrogen concentration in the water showed that the total nitrogen value was affected by N from agricultural land, and the total nitrogen amount was generated from agricultural land and arable land ($p < 0.05$). The higher the concentrations of TN were from arable land and agricultural land, the higher the value of TN was in the water (positive function).

The effect of anthropogenic loads on the ammonium nitrogen concentration (NH_4-N is dependent variable Y) was calculated by multiple regression analysis. The results are presented in Table 3.

Table 2. The influence of anthropogenic loads in rivers basins on the total nitrogen concentration in the water.

Environmental Factor	Unstandardized Coefficients		Standardized Coefficients	t	Significance Level $p < 0.05$
	B	Std. Error	Beta		
Constant	3.853	0.587		6.566	0.000
N from municipal wastewater, t/year	−0.128	0.132	−1.912	−0.969	0.349
N from surface wastewater, t/year	−0.015	0.137	−0.174	−0.108	0.915
N from households not connected to sewage networks, t/year	−0.208	0.239	−1.827	−0.870	0.399
* N from agricultural land, t/year	0.005	0.003	4.745	1.544	0.045
N from background, t/year	−0.011	0.015	−1.736	−0.720	0.484
N from transit pollution, t/year	−0.001	0.001	−0.161	−0.374	0.714
* Agricultural land, ha	0.027	0.017	13.642	1.557	0.042
Forests, ha	0.009	0.017	1.938	0.519	0.612
Wetlands, ha	0.025	0.029	0.268	0.854	0.407
Meadows, ha	0.043	0.047	2.390	0.923	0.372
* Arable land, ha	0.036	0.013	9.809	2.732	0.016
Infertile land, ha	−0.083	0.078	−1.726	−1.063	0.306
Green land, ha	3.010	2.489	2.725	1.210	0.246

Dependent variable: TN; * significance factor, $p < 0.05$.

Table 3. The influence of anthropogenic loads in river basins on the ammonium ion concentration in the water.

Environmental Factor	Unstandardized Coefficients		Standardized Coefficients	t	Significance Level $p < 0.05$
	B	Std. Error	Beta		
Constant	0.033	0.022		1.508	0.154
N from municipal wastewater, t/year	0.006	0.005	1.800	1.203	0.249
N from surface wastewater, t/year	0.005	0.005	1.122	0.921	0.373
* N from households not connected to sewage networks, t/year	0.012	0.009	2.187	1.374	0.049
* N from agricultural land, t/year	0.000	0.000	3.309	1.421	0.047
N from background, t/year	2.408×10^{-5}	0.001	0.080	0.044	0.966
* N from transit pollution, t/year	0.000	0.000	0.852	2.606	0.021
Agricultural land, ha	−0.001	0.001	−5.301	−0.798	0.438
* Forests, ha	−0.002	0.001	−9.231	−3.263	0.006
* Wetlands, ha	0.002	0.001	0.533	2.241	0.042
* Meadows, ha	0.003	0.002	3.674	1.871	0.048
* Arable land, ha	0.001	0.001	5.630	2.069	0.049
* Infertile land, ha	−0.010	0.003	−4.026	−3.271	0.006
* Green land, ha	−0.411	0.093	−7.511	−4.398	0.001

Dependent variable: NH_4-N; * significance factor, $p < 0.05$.

Multiple regression analysis of the influence of anthropogenic loads on the ammonium nitrogen concentration in the water showed that the total NH_4-N value was affected by the TN from households not connected to sewage networks, the TN from agricultural land and transit pollution, and the NH_4-N amount generated from forests, wetlands, meadows, arable land, infertile land, and green land ($p < 0.05$). The higher the concentrations of NH_4-N were from households not connected to the sewage networks, agricultural land, transit pollution, wetlands, meadows, and arable land, the higher the value of the NH_4-N was in the water (positive function). The higher the NH_4-N concentrations were from forests, infertile land, and green land, the lower the NH_4-N concentration was in the water (negative function).

The effect of anthropogenic loads on the nitrate nitrogen concentration (NO_3-N is dependent variable Y) was calculated by multiple regression analysis. The results are presented in Table 4.

Table 4. The influence of anthropogenic loads in basins on the total nitrogen concentration in the water.

Environmental Factor	Unstandardized Coefficients		Standardized Coefficients	t	Significance Level $p < 0.05$
	B	**Std. Error**	**Beta**		
Constant	2.358	0.514		4.584	0.000
N from municipal wastewater, t/year	−0.060	0.116	−1.289	−0.515	0.614
N from surface wastewater t/year	0.006	0.120	0.106	0.052	0.959
N from households not connected to sewage networks, t/year	0.185	0.209	2.348	0.882	0.393
N from agricultural land, t/year	0.001	0.003	1.921	0.493	0.629
N from background; t/year	−0.002	0.013	−0.395	−0.129	0.899
N from transit pollution, t/year	0.000	0.001	0.051	0.094	0.926
Agricultural land, ha	0.013	0.015	9.911	0.893	0.387
Forests, ha	0.005	0.015	1.523	0.322	0.752
Wetlands, ha	0.018	0.026	0.277	0.698	0.497
Meadows, ha	0.023	0.041	1.800	0.549	0.592
* Arable land, ha	0.021	0.012	8.031	1.766	0.049
Infertile land, ha	−0.013	0.068	−0.388	−0.188	0.853
Green land, ha	0.505	2.182	0.661	0.231	0.820

Dependent variable: NO_3-N; * significance factor, $p < 0.05$.

Multiple regression analysis of the influence of anthropogenic loads on the nitrate nitrogen concentration in the water showed that the NO_3-N value was affected only by arable land ($p < 0.05$). The higher the concentration of NO_3-N was from arable land, the higher the value of NO_3-N was in the water (positive function).

The effect of anthropogenic loads on the total phosphorus concentration (TP is dependent variable Y) was calculated by multiple regression analysis. The results are presented in Table 5.

Multiple regression analysis of the influence of anthropogenic loads in basins on the concentration of total phosphorus in the water showed that the total phosphorus value was influenced by the discharge of surface wastewater from households not connected to sewage networks, agricultural land, arable land, infertile land, and green land ($p < 0.05$). The higher the TP concentration was in the surface wastewater from households not connected to sewage networks, agricultural land, arable land, the higher the TP value was in the water (positive function). The larger the infertile and green areas were in the river basins, the lower the total phosphorus concentration was in the water (negative function).

The effect of anthropogenic loads on the phosphate phosphorus concentration, (PO_4-P is dependent variable Y), was calculated by multiple regression analysis. The results are presented in Table 6.

Table 5. The influence of anthropogenic loads in river basins on the total phosphorus concentration in the water.

Environmental Factor	Unstandardized Coefficients		Standardized Coefficients	t	Significance Level $p < 0.05$
	B	Std. Error	Beta		
Constant	0.038	0.007		5.166	0.000
P from municipal wastewater, t/year	0.000	0.001	0.176	0.116	0.910
* P from surface wastewater, t/year	0.013	0.008	1.014	1.688	0.043
* P from households not connected to sewage networks, t/year	0.026	0.011	3.846	2.345	0.034
* P from agricultural land, t/year	0.001	0.001	3.038	1.597	0.043
P from background, t/year	0.000	0.002	0.082	0.072	0.944
P from transit pollution, t/year	−0.002	0.002	−1.034	−1.392	0.186
Agricultural land, ha	9.643×10^{-5}	0.000	4.004	0.453	0.657
Forests, ha	0.000	0.000	3.511	1369	0.192
Wetlands, ha	0.000	0.000	−0.098	−0.298	0.770
Meadows, ha	0.000	0.001	2.101	0.849	0.410
* Arable land, ha	0.000	0.000	9.569	1.963	0.049
* Infertile land, ha	−0.003	0.001	−4.328	−2.552	0.023
* Green land, ha	−0.085	0.038	−6.274	−2.217	0.044

Dependent variable: TP, * significance factor, $p < 0.05$.

Table 6. The influence of anthropogenic loads in river basins on the total phosphorus concentration in the water.

Environmental Factor	Unstandardized Coefficients		Standardized Coefficients	t	Significance Level $p < 0.05$
	B	Std. Error	Beta		
Constant	0.009	0.009		0.981	0.343
* P from municipal wastewater, t/year	0.004	0.001	0.321	3.192	0.007
P from surface wastewater, t/year	0.005	0.010	0.021	0.531	0.604
P from households not connected to sewage networks, t/year	0.006	0.014	0.047	0.429	0.675
P from agricultural land, t/year	0.000	0.001	−0.021	−0.165	0.871
* P from background, t/year	0.004	0.002	0.138	1.827	0.049
* P from transit pollution, t/year	−0.008	0.002	−0.200	−4.061	0.001
* Agricultural land, ha	0.001	0.000	1.732	2.957	0.010
* Forests, ha	−0.000	0.000	−0.350	−2.060	0.049
Wetlands, ha	0.000	0.000	0.006	0.271	0.790
* Meadows, ha	0.002	0.001	0.394	2.402	0.031
* Arable land, ha	0.001	0.000	0.802	2.482	0.026
* Infertile land, ha	−0.006	0.001	−0.550	−4.896	0.000
* Green land, ha	−0.133	0.049	0.514	−2.741	0.016

Dependent variable: PO_4-P, * significance factor, $p < 0.05$.

Multiple regression analysis of the influence of anthropogenic loads in basins on the concentration of phosphate phosphorus in the water showed that the PO_4-P value was influenced by the discharge of municipal wastewater from background and transit pollution, agricultural land, forests, meadows, arable land, infertile land, and green land ($p < 0.05$). The higher the PO_4-P concentration was in the municipal wastewater from background pollution, agricultural land, meadows, arable land, the higher the PO_4-P value was in the water (positive function). The higher the transit pollution, and the larger the forests, infertile, and green areas were in the river basins, the lower the PO_4-P concentration was in the water (negative function).

4. Discussion

Agricultural activity has strict negative impact on condition of surface water bodies, their ecosystems, the degradation of vegetation, and the quantitative and qualitative changes in fish populations in the Mediterranean basin [62]. The main factor affecting the Baltic Sea region environment is the increased amount of nutrients in rivers, mainly from diffuse agricultural sources [63]. The diffused nitrogen of anthropogenic origin account for about 70% of the total load deposited into rivers and lakes of the Baltic Sea basin area. Of the total diffuse load of nitrogen deposited into the Baltic Sea, 80% is from agriculture [64,65]. In Estonia, Latvia, and Lithuania, agriculture was intensified, and the amount of nitrogen fertilizers was increased after the 1990s [66].

Ikauniece and Lagzdinš assessed the status of two rivers, the Slocene and the Age, in Latvia. It was found that the ecological and chemical status of these rivers depended on the following factors: climatic conditions, types of soil and land-use, and human activities. The impact of land-use types and concentrations of total nitrogen, NO_3 -N, NH_4-N, total phosphorus, and $PO_4{}^-$-P on the water of rivers was established. The highest concentrations of these substances were determined in the spring. It can be stated that snow melt during the spring period increases losses of biogenic compounds from concentrated sources [36].

An increase in sensitivity was found in basins with more agricultural land and more fertilizer. A change in the use of chemical fertilizers by $\pm 20\%$ affected the NO_3-N loads in the water body between zero effect and an increase of $\pm 13\%$, while a change in manure use by $\pm 20\%$ affected the NO_3-N loads in the water body from zero effect to a change of -6% to $+7\%$ [63]. Ferrier [67] pointed out that nitrate concentrations in water of the rivers depend on the area of arable land, and there is a relationship between orthophosphate-P, suspended solids concentrations and meadow cover. Studies conducted in the Liswarta basin (Poland) have showed that high concentrations of nutrients in the Liswarta River and its tributaries are closely linked to the agriculture activities in this basin. However, urban wastewater effluents effected the highest concentrations of nutrients set in the Biała Oksza River [34]. Other Polish researchers assessed the influence of land use on the condition of the Dunajec, Czarny Dunajec, Biały Dunajec, and Białka rivers in the Podhale region (southern Poland). The results of their study showed that the concentrated pollution sources, such as effluents from WWTPs or untreated sewage from households, were more important than diffuse sources but agricultural activities significantly affect water quality of rivers [39].

Watershed modelling was used to discover the critical areas of water quality of rivers, as well as to define impacts and identify the most significant pollution sources in the river basins in Lithuania. Regional diffuse pollution leaching patterns were estimated using this model. The largest leaching rates total phosphorus were assessed in the southeastern and western parts of Lithuania. The largest leaching of total nitrogen was determined to occur in the center of the country. It can be seen from the modelling results that agriculture is the dominant pollution source in all Lithuanian river basins. The organic loads from diffuse pollution sources accounted for 60–90% of the annual loads in all of the river basins, excluding the urban catchments of the Neris and Nemunas rivers. The total phosphorus loads from agricultural sources accounted for 50–93% of the annual TP load. The pollution from concentrated sources and non-sewered households had almost no influence on the nitrate loads, and agriculture was the only dominant source of pollution, contributing

90–99% of the annual nitrate load [68]. It was determined that, satisfactorily, 90% of all nitrogen entered the Mūša sub-catchment from the diffuse pollution sources, including 87% from the arable land and just a little more than 3% from the forest territory and pastures. A total of 10% of all nitrogen in the basin came from the concentrated pollution sources. The largest amounts of total phosphorus in the Mūša sub-catchment entered the basin from the concentrated pollution sources (about 49%), arable land (36%) and about 15% from the forest area and pastures [69].

Various sources indicate measures for protection against diffuse pollution. In Poland, the recommendations for the protection of river valleys from biogenic pollution include the activities such as preserving natural vegetation on the banks of rivers, reducing of intensive agriculture activities and others [34]. Scholz [70] introduced diffuse pollution control strategies involving draining the natural wetlands by ditches in Germany.

In Lithuania, the main measures that should be applied to reduce the input of pollution from agricultural activities into rivers and other inland waters are as follows [71]:

- ✓ The application of fertilization plans and targeted/precision farming. Balanced fertilization reduces the need for fertilizers and pesticides and saves water resources. This results in less nitrogen and phosphorus leaching and less eutrophication in surface water bodies;
- ✓ Additional protection strips for surface water bodies. The protective strips of natural vegetation left along the water bodies help to absorb excess nutrients and control water pollution;
- ✓ Stubble fields left during the winter help conserve water resources and prevent nutrient leaching;
- ✓ The installation of controlled drainage. An intelligent drainage system increases yields by reducing the need for fertilizer and stopping the leaching of nutrients into surface water bodies.

5. Conclusions

1. The total nitrogen values did not comply with the requirements of to the "good" and "very good" ecological status classes in 51% of the tested water bodies; 19% had a "bad" to "moderate" BOD_7, 50% had "bad" to "moderate" NH_4-N, 37% had "bad" to "moderate" NO_3-N, 4 % had "bad" to "moderate" PO_4-P, and the total phosphorus values did not correspond to the "good" or "very good" ecological status classes in 4% of the tested water bodies;
2. River basins accumulate the biggest quantities from the following sources: transit pollution, contributing 87,599 t/year of total nitrogen and 5020 t/year of phosphorus; agricultural pollution, contributing 56,031 t/year of total nitrogen and 2474 t/year of total phosphorus;
3. The biggest total nitrogen load in river basins per year is from transit pollution, accounting for 53.89%; agricultural pollution accounts for 34.47%. The highest total phosphorus load is also from transit pollution, accounting for 58.78%; agricultural pollution accounts for 28.97%;
4. The multiple regression analysis showed that the agricultural activity had the biggest negative influence on the ecological status of rivers according to all studied indicators.

Author Contributions: L.Č. designed the study and performed the experiments; L.Č., D.Š. and M.D. performed the experiments, analyzed the data, and wrote the manuscript. All authors have read and agreed to the published version of the manuscript.

Funding: This research received no external funding.

Institutional Review Board Statement: Not applicable.

Informed Consent Statement: Informed consent was obtained from all subjects involved in the study.

Data Availability Statement: The data are available in a publicly accessible repository.

Conflicts of Interest: The authors declare no conflict of interest.

References

1. Aplinkos apsaugos agentūra. Upių, ežerų ir tvenkinių ekologinė būklė/Environmental Protection Agency. Ecological Status of Rivers, Lakes and Ponds. 2017. Available online: https://www.akoalicija.lt/veikla/2019-06-26-d%C4%97l-upi%C5%B3-ekologin%C4%97s-b%C5%ABkl%C4%97s-gerinimo (accessed on 29 August 2021). (In Lithuanian).
2. Tejerina-Garro, F.L.; Maldonado, M.; Ibañez, C.; Pont, D.; Roset, N.; Oberdorff, T. Effects of natural and anthropogenic environmental changes on riverine fish assemblages: A framework for ecological assessment of rivers. *Biol. Appl. Sci. Braz. Arch. Biol. Technol.* **2005**, *48*, 91–108. [CrossRef]
3. Hu, W.; Wang, G.; Deng, W.; Li, S. The influence of dams on ecohydrological conditions in the Huaihe River basin, China. *Ecol. Eng.* **2008**, *33*, 233–241. [CrossRef]
4. Żelazny, M.; Siwek, J.P. Determinants of seasonal changes in streamwater chemistry in small catchments with different land uses: Case study from Poland's Carpathian foothills. *Pol. J. Environ. Stud.* **2012**, *21*, 791–804.
5. Lenart-Boroń, A.; Wolanin, A.; Jelonkiewicz, Ł.; Chmielewska-Błotnicka, D.; Żelazny, M. Spatiotemporal variability in microbiological water quality of the Białka river and its relation to the selected physicochemical parameters of water. *Water Air Soil Pollut.* **2016**, *227*, 22. [CrossRef]
6. Bojarczuk, A.; Jelonkiewicz, L.; Lenart-Boroń, A. The effect of anthropogenic and natural factors on the prevalence of physicochemical parameters of water and bacterial water quality indicators along the river Białka, southern Poland. *Environ. Sci. Pollut. Res.* **2018**, *25*, 10102–10114. [CrossRef]
7. Álvarez, X.; Valero, E.; Torre-Rodríguez, N.; Acuña-Alonso, C. Influence of Small Hydroelectric Power Stations on River Water Quality. *Water* **2020**, *12*, 312. [CrossRef]
8. Zhao, K.; Wu, H.; Chen, W.; Sun, W.; Zhang, H.; Duan, W.; Chen, W.; He, B. Impacts of Landscapes on Water Quality in A Typical Headwater Catchment, Southeastern China. *Sustainability* **2020**, *12*, 721. [CrossRef]
9. Hughes, R.M.; Vadas, R.L., Jr. Agricultural Effects on Streams and Rivers: A Western USA Focus. *Water* **2021**, *13*, 1901. [CrossRef]
10. Burkholder, J.; Libra, B.; Weyer, P.; Heathcote, S.; Kolpin, D.; Thorne, P.S.; Wichman, M. Impacts of waste from concentrated animal feeding operations on water quality. *Environ. Health Perspect.* **2007**, *115*, 308–312. [CrossRef] [PubMed]
11. Páll, E.; Niculae, M.; Kiss, T.; Şandru, C.D.; Spînu, M. Human impact on the microbiological water quality of the rivers. *J. Med. Microbiol.* **2013**, *62*, 1635–1640. [CrossRef]
12. Kroll, S.A.; Oakland, H.C. A review of studies documenting the effects of agricultural best management practices on physiochemical and biological measures of stream ecosystem integrity. *Nat. Areas J.* **2019**, *39*, 58–77. [CrossRef]
13. Liu, J.; Han, G.; Liu, X.; Liu, M.; Song, C.; Zhang, Q.; Yang, K.; Li, X. Impacts of Anthropogenic Changes on the Mun River Water: Insight from Spatio-Distributions and Relationship of C and N Species in Northeast Thailand. *Int. J. Environ. Res. Public Health* **2019**, *16*, 659. [CrossRef]
14. Liu, Y.; Li, H.; Cui, G.; Cao, Y. Water quality attribution and simulation of non-point source pollution load flux in the Hulan River basin. *Sci. Rep.* **2020**, *10*, 3012. [CrossRef] [PubMed]
15. Castillo, M.M.; Allan, J.D.; Brunzell, S. Nutrient concentrations and discharges in a Midwestern agricultural catchment. *J. Env. Qual.* **2000**, *29*, 1142–1151. [CrossRef]
16. Valiela, I.; Bowen, J.L. Nitrogen sources to watersheds and estuaries: Role of land cover mosaics and losses within watersheds. *Environ. Pollut.* **2002**, *118*, 239–248. [CrossRef]
17. Buck, O.; Niyogi, D.K.; Townsend, C.R. Scale-dependence of land use effects on water quality of streams in agricultural catchments. *Environ. Pollut.* **2004**, *130*, 287–299. [CrossRef] [PubMed]
18. Alam, M.N.; Elahi, F.; Alam, D.U. Risk and water quality assessment overview of River Sitalakhya in Bangladesh. *Acad. Open Internet J.* **2006**, *19*, 112–202.
19. Zhang, Q.L.; Shi, X.Z.; Huang, B.; Yu, D.S.; Öbornc, I.; Blombäckc, K.; Wang, H.J.; Pagella, T.F.; Sinclair, F.L. Surface water quality of factory-based and vegetable-based peri-urban areas in the Yangtze River delta region, China. *Catena* **2007**, *69*, 57–64. [CrossRef]
20. Hussain, M.; Ahmed, S.M.; Abderrahman, W. Cluster analysis and quality assessment of logged water at an irrigation project, eastern Saudi Arabia. *J. Environ. Manag.* **2008**, *86*, 297–307. [CrossRef]
21. Khatri, N.; Tyagi, S. Influences of natural and anthropogenic factors on surface and groundwater quality in rural and urban areas. *Front. Life Sci.* **2015**, *8*, 23–39. [CrossRef]
22. Allan, J.D. Landscapes and riverscapes: The influence of land use on stream ecosystems. *Annu. Rev. Ecol. Evol. Syst.* **2004**, *35*, 257–284. [CrossRef]
23. Bhat, S.; Jacobs, J.M.; Hatfield, K.; Prenger, J. Relationships between stream water chemistry and military land use in forested watersheds in Fort Benning. *Georgia. Ecol. Ind.* **2006**, *6*, 458–466. [CrossRef]
24. Hopkins, R.L. Use of landscape pattern metrics and multiscale data in aquatic species distribution models: A case study of a freshwater mussel. *Landsc. Ecol.* **2009**, *24*, 943–955. [CrossRef]
25. Tran, C.P.; Bode, R.W.; Smith, A.J.; Kleppel, G.S. Land-use proximity as a basis for assessing stream water quality in New York State (USA). *Ecol. Indic.* **2010**, *10*, 727–733. [CrossRef]
26. Weijters, M.J.; Janse, J.H.; Alkemade, R.; Verhoeven, J.T.A. Quantifying the effect of catchment land use and water nutrient concentrations on freshwater river and stream biodiversity. *Aquat. Conserv. Mar. Freshw. Ecosyst.* **2009**, *19*, 104–112. [CrossRef]

27. Hollister, J.W.; August, P.V.; Paul, J.F. Effects of spatial extent on landscape structure and sediment metal concentration relationships in small estuarine systems of the United States' Mid-Atlantic Coast. *Landsc. Ecol.* **2008**, *23*, 91–106. [CrossRef]

28. Sliva, L.; Williams, D.D. Buffer zone versus whole catchment approaches to studying land use impact on river water quality. *Water Res.* **2001**, *35*, 3462–3472. [CrossRef]

29. Brabec, E.A. Imperviousness and land-use policy: Toward an effective approach to watershed planning. *J. Hydrol. Eng.* **2009**, *14*, 425–433. [CrossRef]

30. Lee, S.W.; Hwang, S.J.; Lee, S.B.; Hwang, H.S.; Sung, H.C. Landscape ecological approach to the relationships of land use patterns in watersheds to water quality characteristics. *Landsc. Urban Plan.* **2009**, *92*, 80–89. [CrossRef]

31. Kelly, M.; Bennion, H.; Burgess, A.; Ellis, J.; Juggins, S.; Guthrie, R.; Jamieson, J.; Adriaenssens, V.; Yallop, M. Uncertainty in ecological status assessments of lakes and rivers using diatoms. *Hydrobiologia* **2009**, *633*, 5–15. [CrossRef]

32. Nõges, P.; Van de Bund, W.; Cardoso, A.K.; Solimini, A.G.; Heiskanen, A.S. Assessment of the ecological status of European surface waters: A work in progress. *Hydrobiologia* **2009**, *633*, 197–211. [CrossRef]

33. Ayobahan, S.U.; Ezenwa, I.M.; Orogun, E.E.; Uriri, J.E.; Wemimo, I.J. Assessment of Anthropogenic Activities on Water Quality of Benin River. *J. Appl. Sci. Environ. Manag.* **2014**, *18*, 629–636. [CrossRef]

34. Matysik, M.; Atsalon, D.; Ruman, M. Surface Water Quality in Relation to Land Cover in Agricultural Catchments (Liswarta River Basin Case Study). *Pol. J. Environ. Stud.* **2015**, *24*, 175–184. [CrossRef]

35. Česonienė, L.; Šileikienė, D.; Dapkienė, M.; Radzevičius, A.; Räsänen, K. Assessment of chemical and microbiological parameters on the Leite River Lithuania. *Environ. Sci. Pollut. Res.* **2019**, *26*, 18752–18765. [CrossRef] [PubMed]

36. Ikauniece, K.; Lagzdinš, A. The assessment of chemical and ecological status in the water bodies of Slocene and Age. *Water Manag. Res. Rural Dev.* **2020**, *35*, 241–247. [CrossRef]

37. Grizzetti, B.; Pistocchi, A.; Liquete, C.; Udias, A.; Bouraoui, F.; van de Bund, W. Human pressures and ecological status of European rivers. *Sci. Rep.* **2017**, *7*, 205. [CrossRef] [PubMed]

38. Ruminaitė, R. Research and Evaluation of the Anthropogenic Activity Impact on the River Runoff and Water Quality. Ph.D. Thesis, Vilnius Gediminas Technical University, Vilnius, Lithuania, 2010. Available online: http://vddb.laba.lt/fedora/get/LT-eLABa-0001:E.02~{}2010~{}D_20101222_130641-59851/DS.005.0.01.ETD (accessed on 5 May 2021). (In Lithuanian).

39. Lenart-Boroń, A.; Wolanin, A.; Jelonkiewicz, E.; Żelazny, M. The effect of anthropogenic pressure shown by microbiological and chemical water quality indicators on the main rivers of Podhale, southern Poland. *Env. Sci. Pollut. Res. Int.* **2017**, *24*, 12938–12948. [CrossRef] [PubMed]

40. Cesoniene, L.; Dapkiene, M.; Sileikiene, D.; Rekasiene, V. Impact of Wastewater Treatment Plant on Water Quality of the River Mazoji Sruoja, Plunge District. *J. Environ. Res. Eng. Manag.* **2017**, *73*, 33–44.

41. Joshua, N.; Edokpayi, J.N.; Odiyo, J.O.; Durowoju, O.S. Impact of Wastewater on Surface Water Quality in Developing Countries: A Case Study of South Africa. In *Water Quality*; IntechOpen: London, UK, 2017; pp. 401–416. Available online: https://www.intechopen.com/chapters/53194 (accessed on 5 May 2021). [CrossRef]

42. Alcon, F.; de-Miguel, M.D.; Martínez-Paz, J.M. Assessment of real and perceived cost-effectiveness to inform agricultural diffuse pollution mitigation policies. *Land Use Policy* **2021**, *107*, 104561. [CrossRef]

43. Matono, P.; Batista, T.; Sampaio, E.; Ilhéu, M. Effects of Agricultural Land Use on the Ecohydrology of Small-Medium Mediterranean River Basins: Insights from a Case Study in the South of Portugal. In *Land Use—Assessing the Past, Envisioning the Future*; IntechOpen: London, UK, 2019; pp. 29–51.

44. Liu, J.; Shen, Z.; Chen, L. Assessing how spatial variations of land use pattern affect water quality across a typical urbanized watershed in Beijing, China. *Landsc. Urban Plan.* **2018**, *176*, 51–63. [CrossRef]

45. Sveikauskaite, I. Lithuanian River Pollution and Water Quality Trends in 1992–2009. Master's Thesis, Vytautas Magnus University, Kaunas, Lithuania, 2011.

46. Fogelberg, S. Modelling nitrogen retention at the catchment scale.Report UPTEC W 03 019. Uppsala: Uppsala University. P. 1-51.French urban catchment. *Water Res.* **2003**, *85*, 432–442.

47. Organisation for Economic Co-operation and Development. *Environmental Performance of Agriculture in OECD Countries Since1990*; OECD: Paris, France, 2008; ISBN 9789264040922.

48. Kyllmar, K.; Bechmann, M.; Deelstra, J.; Iital, A.; Blicher-Mathiesen, G.; Jansons, V.; Koskiaho, J.; Povilaitis, A. Long-term monitoringof nutrient losses from agricultural catchments in the Nordic-Baltic region-A discussion of methods, uncertainties. *Agric. Ecosyst. Environ.* **2014**, *198*, 4–12. [CrossRef]

49. Povilaitis, A.; Šileika, A.; Deelstra, J.; Gaigalis, K.; Baigys, G. Nitrogen losses from small agricultural catchments in Lithuania. *Agric. Ecosyst. Environ.* **2014**, *198*, 54–64. [CrossRef]

50. Xiao, L.; Liu, J.; Ge, J. Dynamic game in agriculture and industry cross-sectoral water pollution governance in developing countries. *Agric. Water Manag.* **2021**, *243*, 106417. [CrossRef]

51. Waller, D.M.; Meyer, A.G.; Raff, Z.; Apfelbaum, S.I. Shifts in precipitation and agricultural intensity increase phosphorus concentrations and loads in an agricultural watershed. *J. Environ. Manag.* **2021**, *284*, 112019. [CrossRef] [PubMed]

52. USEPA (U.S. Environmental Protection Agency). National Water Quality Inventory Report. EPA-841-F-02-003. 2002. Available online: https://www.epa.gov/sites/production/files/2015-09/documents/2007_10_15_305b_2002report_report2002305b.pdf. (accessed on 30 August 2021).

53. Staniszewska, M.; Schnug, E. Status of organic agriculture in thecountries of the Baltic Sea region. *Landbauforsch. Volkenrode* **2002**, *52*, 75–80.
54. Vincevičienė, V.; Asadauskaitė, A. Assessment of Point Pollution and Impact to Water Quality of the Transboundary River Basins–the Venta and the Musa–Lielupe. *Environ. Res. Eng. Manag.* **2002**, *1*, 41–51.
55. Ruminaitė, R. Water runoff and pollution peculiarities in Lielupe watershed per 1992–2006-year period. *Water Manag. Eng.* **2010**, *36*, 100–108.
56. Povilaitis, A. Source apportionment and retention of nutrients and organic matter in the merkys river basin in southern Lithuania. *J. Environ. Eng. Landsc. Manag.* **2008**, *16*, 195–204. [CrossRef]
57. Plungė, S.; Gudas, M. Žemės ūkis ir Lietuvos vandenys. Žemės ūkio veiklos poveikis Lietuvos upių būklei ir taršos apkrovoms į Baltijos jūrą/Impact of Agricultural Activities on the Condition of Lithuanian Rivers and Pollution Loads in the Baltic Sea. 2018. Available online: https://www.akoalicija.lt/veikla/2019-06-26-d%C4%97l-upi%C5%B3-ekologin%C4%97s-b%C5%ABkl%C4%97s-gerinimo (accessed on 29 August 2021). (In Lithuanian).
58. Povilaitis, A. Nutrient Retention in Surface Waters of Lithuania. *Pol. J. Environ. Stud.* **2011**, *20*, 1575–1584.
59. Plungė, S.; Gudas, M. Klimato kaitos poveikis vandens telkiniams. Priemonių taršai iš žemės ūkio šaltinių sumažinti galimybių įvertinimas klimato kaitos perspektyvoje/Impact of Climate Change on Water Bodies. Assessment of Measures to Reduce Pollution from Agricultural Sources in the Perspective of Climate Change. 2021. Available online: https://vanduo.old.gamta.lt/files/report1627991771354.html (accessed on 29 August 2021).
60. Approval of the Description of the Procedure for Assessing the Ecological Status of Surface Water Bodies. Order of the Minister of the Environment of the Republic of Lithuania: 12 April 2007 No. D1-210/Lietuvos Respublikos Aplinkos Ministro Įsakymas dėl Paviršinių Vandens Telkinių Būklės Nustatymo Metodikos Patvirtinimo 2007m. Vilnius. Balandzio 12 d. Nr. D1; 2017. Available online: https://portalcris.vdu.lt/cris/bitstream/20.500.12259/115090/1/tomas_kantaravicius_md.pdf (accessed on 5 May 2021).
61. Arnold, J.G.; Srinisvan, R.; Muttiah, R.S.; Williams, J.R. Large area hydrologic modeling and assessment. Part I: Model development. *J. Am. Water Resour.* **1998**, *34*, 73–89. [CrossRef]
62. Hermoso, V.; Clavero, M.; Kennard, M.J. Determinants of fine-scale homogenization and differentiation of native freshwater fish faunas in a Mediterranean Basin: Implications for conservation. *Divers. Distrib.* **2012**, *18*, 236–247. [CrossRef]
63. Thodsen, H.; Farkas, C.; Chormanski, J.; Trolle, D.; Blicher-Mathiesen, G.; Grant, R.; Engebretsen, A.; Kardel, I.; Andersen, H.E. Modelling Nutrient Load Changes from Fertilizer Application Scenarios in Six Catchments around the Baltic Sea. *Agriculture* **2017**, *7*, 41. [CrossRef]
64. HELCOM. Eutrophication in the Baltic Sea—An integrated thematic assessment of the effects of nutrient enrichment and eutrophication in the Baltic Sea region. *Balt. Sea Environ. Proc.* **2009**, *115*, 148.
65. Voss, M.; Dippner, J.W.; Humborg, C.; Hürdler, J.; Korth, F.; Neumann, T.; Schernewski, G.; Venohr, M. History and scenarios of future development of Baltic Sea eutrophication. *Estuar. Coast. Shelf Sci. J.* **2011**, *92*, 307–322. [CrossRef]
66. Andersen, H.E.; Blicher-Mathiesen, G.; Bechmann, M.; Povilaitis, A.; Iital, A.; Lagzdins, A.; Kyllmarf, K. Reprint of "Mitigating diffuse nitrogen losses in the Nordic-Baltic countries". *Agric. Ecosyst. Environ.* **2014**, *198*, 127–134. [CrossRef]
67. Ferrier, R.C.; Edwards, A.C.; Hirst, D.; Littlewood, I.G.; Watts, C.D.; Morris, R. Water quality of Scottish rivers: Spatial and temporal trends. *Sci. Total Environ.* **2001**, *265*, 327–342. [CrossRef]
68. Vaitiekūnienė, J.; Hansen, F.T. National scale water modelling to assess and predict pollution trends in Lithuanian rivers. *Environ. Res. Eng. Manag.* **2005**, *4*, 30–42.
69. Stankevičienė, R. Research and Evaluation of the Anthropogenic Activity Impact on the River Runoff and Water Quality. Summary of Doctoral Dissertation; Vilnius Gediminas Technical University (VILNIUS TECH); Vilnius Technica. 2010. Available online: https://talpykla.elaba.lt/elaba-fedora/objects/elaba:1862147/datastreams/MAIN/content (accessed on 12 April 2021).
70. Scholz, M. *Wetlands for Water Pollution Control*, 2nd ed.; Elsevier Science: Saldorf, UK, 2016.
71. Rural Development. Official Document on Improving the Ecological Condition of Lithuanian Rivers and Other Water Bodies. 2019. Available online: https://www.akoalicija.lt/veikla/2019-06-26-d%C4%97l-upi%C5%B3-ekologin%C4%97s-b%C5%ABkl%C4%97s-gerinimo (accessed on 29 August 2021).

Article

Attribution Analysis of Climate and Anthropic Factors on Runoff and Vegetation Changes in the Source Area of the Yangtze River from 1982 to 2016

Guangxing Ji [1], Huiyun Song [1], Hejie Wei [1] and Leying Wu [2,*]

[1] College of Resources and Environmental Sciences, Henan Agricultural University, Zhengzhou 450002, China; guangxingji@henau.edu.cn (G.J.); HuiyunSong@stu.henau.edu.cn (H.S.); hjwei@henau.edu.cn (H.W.)

[2] Key Research Institute of Yellow River Civilization and Sustainable Development & Collaborative Innovation Center on Yellow River Civilization jointly built by Henan Province and Ministry of Education, Henan University, Kaifeng 475001, China

* Correspondence: 40150007@vip.henu.edu.cn; Tel.: +86-153-1780-1989

Abstract: Analyzing the temporal variation of runoff and vegetation and quantifying the impact of anthropic factors and climate change on vegetation and runoff variation in the source area of the Yangtze River (SAYR), is of great significance for the scientific response to the ecological protection of the region. Therefore, the Budyko hypothesis method and multiple linear regression method were used to quantitatively calculate the contribution rates of climate change and anthropic factors to runoff and vegetation change in the SAYR. It was found that: (1) The runoff, NDVI, precipitation, and potential evaporation in the SAYR from 1982 to 2016 all showed an increasing trend. (2) The mutation year of runoff data from 1982 to 2016 in the SAYR is 2004, and the mutation year of NDVI data from 1982 to 2016 in the SAYR is 1998. (3) The contribution rates of precipitation, potential evaporation and anthropic factors to runoff change of the SAYR are 75.98%, −9.35%, and 33.37%, respectively. (4) The contribution rates of climatic factors and anthropic factors to vegetation change of the SAYR are 38.56% and 61.44%, respectively.

Keywords: runoff variation; vegetation change; attribution analysis; source region of the Yangtze River

Citation: Ji, G.; Song, H.; Wei, H.; Wu, L. Attribution Analysis of Climate and Anthropic Factors on Runoff and Vegetation Changes in the Source Area of the Yangtze River from 1982 to 2016. *Land* **2021**, *10*, 612. https://doi.org/10.3390/land10060612

Academic Editors: Dawid Bedla, Marek Ryczek, Artur Radecki-Pawlik and Wiktor Halecki

Received: 6 May 2021
Accepted: 4 June 2021
Published: 8 June 2021

Publisher's Note: MDPI stays neutral with regard to jurisdictional claims in published maps and institutional affiliations.

Copyright: © 2021 by the authors. Licensee MDPI, Basel, Switzerland. This article is an open access article distributed under the terms and conditions of the Creative Commons Attribution (CC BY) license (https:// creativecommons.org/licenses/by/ 4.0/).

1. Introduction

The source area of the Yangtze River (SAYR) is located in the hinterland of the Qinghai Tibet Plateau, which is an important ecological barrier in China [1]. It has the ecological functions of maintaining the ecological security of the Yangtze River Basin [2]. At the same time, it is also one of the most sensitive and fragile regions in the global ecological environment [3,4]. Due to the joint impact of climate change and anthropic factors (indiscriminating felling of trees, disafforestation, overgrazing, mining, and digging), the ecological environment of the SAYR is deteriorating, and the function of water conservation is weakening [5,6]. Additionally, the problems of grassland degradation, wetland shrinkage, glacier retreat, soil erosion, and biodiversity decline are becoming increasingly prominent. These issues highlight the threats to the water resources and ecological security of the whole basin.

The impact of climate change has been embodied in the components of the hydrological cycle, including precipitation [7], evapotranspiration [8], and runoff [9]. Most of the literature has focused on precipitation and evapotranspiration worldwide [10,11], and runoff has taken less attention. Nevertheless, runoff remains one of the most important water resources and has a great influence on the formation of geomorphology, the development of soil, and the growth of plants [12]. It is an important condition for the regional industrial and agricultural water supply and plays an important role in the development of social economy.

In recent years, under the joint influence of climate change and anthropic factors, the problems of climate, ecology, and hydrology in the SAYR have been a cause for concern. Therefore, numerous programs have been carried out to improve the eco-hydrological environment, among which Grain for Green proposed by the Chinese government in 1998 is the most well-known. In this context, monitoring the dynamic changes of runoff and vegetation and assessing the impacts of anthropic factors and climate change on runoff and vegetation change are of great practical value for formulating rational adaptive management countermeasures [13–19].

Previous studies have focused on the relationship between runoff and vegetation change and its influencing factors in the SAYR. Chen et al. analyzed the temporal and spatial change of vegetation cover in the SAYR from 1982 to 2003 and analyzed the influence of topography and human factors on the vegetation change [20]. Yao et al. analyzed the spatiotemporal variation characteristics of vegetation net primary productivity in the SAYR from 1959 to 2008 and analyzed the relationship between vegetation change and climate factors (temperature, relative humidity, precipitation, wind speed, and sunshine hours) [21]. Liu et al. analyzed the spatiotemporal variation characteristics of vegetation coverage in the SAYR from 1997 to 2012 and believed that the improvement of vegetation status in the SAYR benefited from the joint influences of climate humidification and ecological engineering [22]. Qian et al., Li et al., and Luo et al. studied the variation regularities of runoff in the source area from 1957 to 2009, 1961 to 2011, and 1961 to 2016, respectively, and pointed out that the annual surface water resources in the SAYR all showed an increasing trend [23–25]. Xu et al. analyzed the runoff variation characteristics and the degree of human impact in the SAYR from 1956 to 2004 [26]. Liu et al. separated the influences of climate and anthropic factors on runoff in the SAYR based on the Budyko hypothesis [27]. These research results have important scientific value for understanding the regularities of runoff and vegetation change in the SAYR, but there are few studies on the assessment and quantitative analysis of climate change and the contribution rate of anthropic factors to runoff and vegetation change in the SAYR [27,28].

The objective of this study was to quantify the impact of anthropic factors and climate change on vegetation and runoff variation in the SAYR by following three steps: (1) analyzing the temporal variation regularities of runoff and vegetation in the SAYR; (2) distinguishing the mutation year of runoff and NDVI data with the Mann–Kendall mutation analysis method; and (3) quantifying the contribution rate of climatic and anthropic factors to vegetation and runoff in the SAYR.

2. Study Area and Data

2.1. Study Area

The source region of the Yangtze River is located between 90°43′ and 96°45′ E, 32°30′ and 35°35′ N, with an area of about 13.77×10^4 km² (Figure 1). Its annual average runoff is about 129.17×10^8 m³, and it affects the interannual fluctuation, long-term evolution trend, and the sustainable utilization of water resources in the Yangtze River Basin. The landform is mainly high plains and hills, with an average altitude of more than 4000 m. Its climate type belongs to the semi-humid and semi-arid region of the plateau frigid zone, with abundant sunshine and large temperature difference between day and night. Temperatures are generally low throughout the year in the Yangtze River source area. The month with the lowest precipitation is December, the month with the highest precipitation is July, and the annual precipitation is concentrated in May–September, which accounts for more than 85% of the annual precipitation [29].

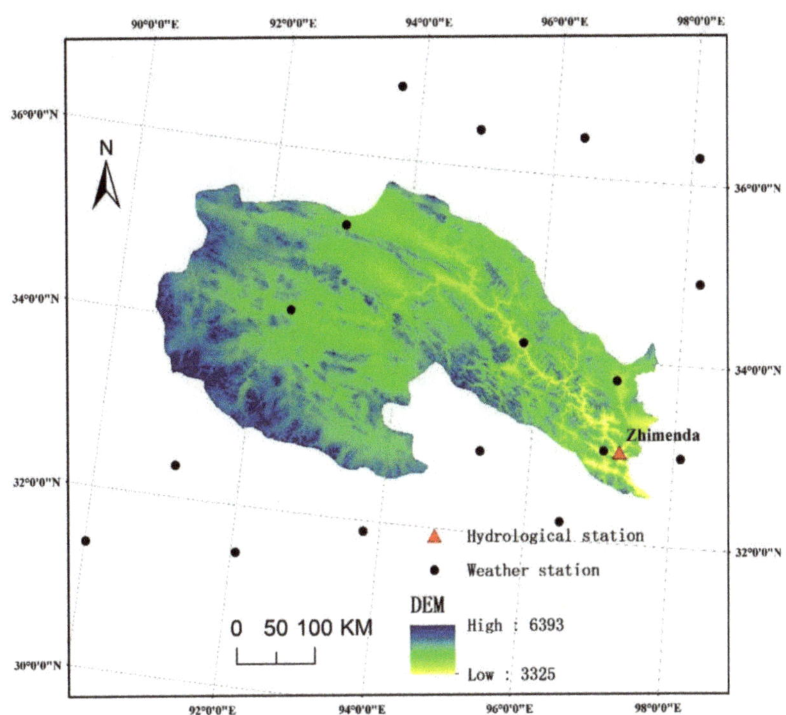

Figure 1. The location of hydrological and meteorological stations in and around the source region of the Yangtze River Basin.

2.2. Data Sources

The data used in this study consist of three parts: (1) The annual runoff observation data of Zhimenda station from 1982 to 2016 were all from the Yangtze River Water Conservancy Commission (http://www.cjw.gov.cn/, accessed on 1 January 2021); (2) the climate station data in and around the source region of the Yangtze River Basin from 1982 to 2016 were obtained from the China Meteorological Administration (http://www.cma.gov.cn, accessed on 1 January 2021). First, the daily potential evaporation of 25 meteorological stations was calculated using the Penman–Monteith equation, and then the monthly precipitation and potential evaporation of 25 meteorological stations were calculated. Finally, the monthly precipitation and potential evaporation were interpolated from data of 25 meteorological stations in and around the SRYR by kriging. Annual precipitation and potential evaporation were obtained by adding monthly scale data. (3) The NDVI data of the source region of the Yangtze River Basin from 1982 to 2016 were obtained from the NOAA CDR AVHRR NDVI: Normalized Difference Vegetation Index, Version 5 (https://www.ncdc.noaa.gov/cdr/terrestrial/normalized-difference-vegetation-index, accessed on 1 January 2021). The time resolution is daily, and the spatial resolution is $0.05° \times 0.05°$. The monthly NDVI and annual NDVI were extracted with the maximum combination method.

3. Research Methods

3.1. Trend Analysis Method

In this study, the univariate linear regression method (*Slope*) was used to analyze the variation trend of runoff, precipitation, potential evaporation, and NDVI in the study period; time t was taken as the independent variable. The calculation equation is as follows [30–32]:

$$Slope = \frac{n \sum_{i=1}^{n} iX_i - \sum_{i=1}^{n} i \sum_{i=1}^{n} X_i}{n \sum_{i=1}^{n} i^2 - \left(\sum_{i=1}^{n} i\right)^2} \tag{1}$$

where *Slope* is the simple linear regression coefficient of runoff, precipitation, potential evaporation, and NDVI from 1982 to 2016. If *Slope* is positive, it means that the variable has an upward trend in the study period, whereas a negative value represents a downward trend; n is the number of years of the study period, i is 1 to n, n is 35, and X represents the annual runoff, precipitation, potential evaporation, and NDVI in the study area.

3.2. Mann–Kendall Mutation Analysis Method

Mann–Kendall (MK) mutation analysis has been widely used to test the abrupt point of hydro-meteorological elements due to its advantages of minimized interference from outliers and simple calculation [33,34].

For X with n data, a rank sequence is constructed, as shown in Equations (2) and (3):

$$Sk = \sum_{i=1}^{k} ri \ (k = 2, 3, \ldots, i) \tag{2}$$

$$ri = \begin{cases} +1 & xi > xj \\ 0 & xi \leq xj \end{cases} (j = 2, 3, \ldots, i) \tag{3}$$

The UF_k statistics can be calculated by Equation (4):

$$UF_k = \frac{[sk - E(sk)]}{\sqrt{Var(sk)}} (k = 1, 2, \ldots, n) \tag{4}$$

where $UF_1 = 0$, $E(sk)$ and $Var(sk)$ represent the mean and variance of sk, respectively. The calculation equations are as follows:

$$E(sk) = \frac{n(n+1)}{4} \ (2 \leq k \leq n) \tag{5}$$

$$Var(sk) = \frac{n(n-1)(2n+5)}{72} \ (2 \leq k \leq n) \tag{6}$$

The inverse order of X is calculated again, and at the same time, $UB_k = UF_k$ ($k = n$, $n - 1, \ldots ,1$), $UB_1 = 0$, then the UB_k curve can be calculated. When the two lines (UF curve and UB curve) intersect, and the intersection point is within the range of the 0.05 significance level, the intersection point can be considered as the mutation year.

3.3. Budyko Hypothesis

There are three assumptions in the attribution analysis of runoff change using the Budyko hypothesis: (1) Human activities and climate change do not affect each other and are independent factors. (2) For multi-year water balance, the change of water storage is usually negligible compared with the average annual precipitation depth. Therefore, the change of catchment storage water for multi-year water balance is assumed as 0; (3) The base period is only affected by climate change. Therefore, except for climate change, other factors affecting runoff change are classified as human activities.

The water balance equation of the basin is expressed as follows:

$$R = P - ET - \Delta S \tag{7}$$

R is the runoff depth; P is precipitation; ET is the actual evapotranspiration; and ΔS is the change of water storage. For multi-year water balance, the change of water storage is usually negligible compared with the average annual precipitation depth. Therefore, in the study of long-term hydrological data, it is generally considered that ΔS is 0.

The Budyko hypothesis, based on an assumption that the change of catchment storage water for multi-year water balance is considered as 0, is widely applied to quantitatively calculate the impact of climate change and anthropic factors on long-term annual runoff change [35–39]. *ET* can be calculated according to the Budyko hypothesis, the calculation equation of which is as follows [40,41]:

$$ET = \frac{P \times ET0}{(P^\omega + ET0^\omega)^{1/\omega}} \tag{8}$$

ET_0 is the annual average potential evapotranspiration (mm); ω indicates characteristic parameters of underlying surface.

ET_0 can be calculated using the Penman–Monteith equation.

$$ET_0 = \frac{0.408\Delta(R_n - G) + \gamma\frac{900}{T+273}U_2(e_a - e_d)}{\Delta + \gamma(1 + 0.34U_2)} \tag{9}$$

Equation (7) can be transformed into Equation (10):

$$R = P - \frac{P \times ET0}{(P^\omega + ET0^\omega)^{1/\omega}} \tag{10}$$

where R, P, and ET_0 are known, and parameter ω can be calculated by the "fsolve" function of MATLAB.

Based on Equation (10), the elastic coefficients of precipitation (ε_P), potential evaporation (ε_{ET0}), and underlying surface parameters (ε_ω) to runoff can be calculated by the following equations [42]. The concept of elastic coefficient is shown in Appendix A.

$$\varepsilon_P = \frac{(1 + \phi^\omega)^{1/\omega+1} - \phi^{\omega+1}}{(1 + \phi^\omega)\left[(1 + \phi^{-\omega})^{1/\omega} - \phi\right]} \tag{11}$$

$$\varepsilon_{ET0} = \frac{1}{(1 + \phi^\omega)\left[1 - (1 + \phi^{-\omega})^{1/\omega}\right]} \tag{12}$$

$$\varepsilon_\omega = \frac{\ln(1 + \phi^\omega) + \phi^\omega \ln(1 + \phi^\omega)}{\omega(1 + \phi^\omega)\left[1 - (1 + \phi^{-\omega})^{1/\omega}\right]} \tag{13}$$

$$\phi = \frac{ET0}{P} \tag{14}$$

The runoff variation amount caused by the annual average precipitation change (ΔRP), annual average potential evaporation change ($\Delta RET0$), and underlying surface parameter change ($\Delta R\omega$) can be calculated by the following Equations (15)–(17):

$$\Delta RP = \varepsilon P \frac{R}{P} \times \Delta P \tag{15}$$

$$\Delta RET0 = \varepsilon ET0 \frac{R}{ET0} \times \Delta ET0 \tag{16}$$

$$\Delta R\omega = \varepsilon\omega \frac{R}{\omega} \times \Delta\omega \tag{17}$$

On this basis, the contribution rate of precipitation (ηR_p), potential evaporation, (ηR_{ET0}) and anthropic factor (ηR_H) to runoff changes can be calculated using Equations (19)–(21):

$$\Delta R = \Delta RP + \Delta RET0 + \Delta R\omega \tag{18}$$

$$\eta R_P = \Delta RP / \Delta R \times 100\% \tag{19}$$

$$\eta R_{ET0} = \Delta RET0 / \Delta R \times 100\% \tag{20}$$

$$\eta R_H = \Delta R\omega / \Delta R \times 100\% \tag{21}$$

3.4. Attribution Analysis of Climate and Anthropic Factors to Vegetation Change

There are two assumptions in the attribution analysis of vegetation change using this method: (1) Human activities and climate change do not affect each other and are independent factors. (2) The base period is only affected by climate change. Therefore, except for climate change, other factors affecting vegetation change are classified as human activities.

According to the mutation analysis results of NDVI time series data, the NDVI data are divided into two parts: base period (T_1) and change period (T_2). On this basis, the change of average NDVI in the two periods ($\triangle \overline{NDVI}$) can be calculated as follows:

$$\triangle \overline{NDVI} = \overline{NDVI_{T2}} - \overline{NDVI_{T1}} \tag{22}$$

where $\overline{NDVI_{T1}}$ and $\overline{NDVI_{T2}}$ are the average NDVI values in the base period (T_1) and change period (T_2), respectively. This method assumes that NDVI in the base period is only affected by climate change. Therefore, the NDVI difference between the base period and change period can be attributed to climatic factors ($\triangle \overline{NDVI_C}$) and anthropogenic factors ($\triangle \overline{NDVI_H}$).

The monthly NDVI values are quantitatively correlated with monthly precipitation and potential evapotranspiration [43,44]. Therefore, a multiple linear regression equation between monthly NDVI, monthly precipitation, and monthly potential evapotranspiration in the base period (T_1) can be established:

$$\overline{NDVI_{T1}} = a * PT1 + b * ET0T1 + c \tag{23}$$

The monthly precipitation (P_{T2}) and potential evapotranspiration ($ET0_{T2}$) in the T_2 period are known, and the simulated NDVI in the T_2 period ($\overline{NDVI_{T2,S}}$) can be calculated by Equation (24).

$$\overline{NDVI_{T2,S}} = a * PT2 + b * ET0T2 + c \tag{24}$$

This method assumes that NDVI in the base period ($\overline{NDVI_{T1}}$) is only affected by climate change. Therefore, $\overline{NDVI_{T2,S}}$ in the T_2 period calculated by Equation (24) is only affected by climate change. $\overline{NDVI_{T2}}$ is the average value of NDVI observation in the T_2 period, which is affected by the combination of climate change and human activities.

The contribution rates of anthropic factor ($\eta NDVI_H$) and climate change ($\eta NDVI_C$) to NDVI can be calculated separately by Equations (25)–(28):

$$\triangle \overline{NDVI_H} = \overline{NDVI_{T2}} - \overline{NDVI_{T2,S}} \tag{25}$$

$$\triangle \overline{NDVI_C} = \overline{NDVI_{T2,S}} - \overline{NDVI_{T1}} \tag{26}$$

$$\eta NDVI_H = \triangle \overline{NDVI_H} / \triangle \overline{NDVI} \tag{27}$$

$$\eta NDVI_C = \triangle \overline{NDVI_C} / \triangle \overline{NDVI} \tag{28}$$

4. Results and Analysis

4.1. Trends Analysis of Runoff, NDVI, and Climate Factors

Figure 2 displays the change trend of the annual runoff in Zhimenda hydrological station and NDVI in the SAYR. It can be found that the annual runoff of Zhimenda hydrological station shows an increasing trend from 1982 to 2018, with an average annual increase of 0.8831×10^8 m^3. The NDVI value in the SAYR showed a significant growth trend from 1982 to 2016, with an average annual growth of 0.0018.

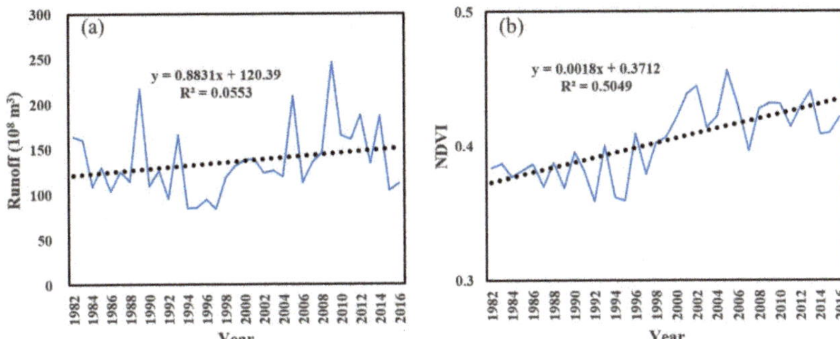

Figure 2. Change trend of annual runoff (**a**) at Zhimenda hydrological station and NDVI (**b**) in the SAYR.

Figure 3 displays the change trend of average annual precipitation and annual potential evaporation in the SAYR. It can be found from Figure 3a that the annual average precipitation in the SAYR showed an increasing trend from 1982 to 2016, with an average annual increase of 1.2394 mm. Figure 3b displays that the average annual potential evaporation in the SAYR also showed an upward trend from 1982 to 2016, with an average annual increase of 0.5965 mm.

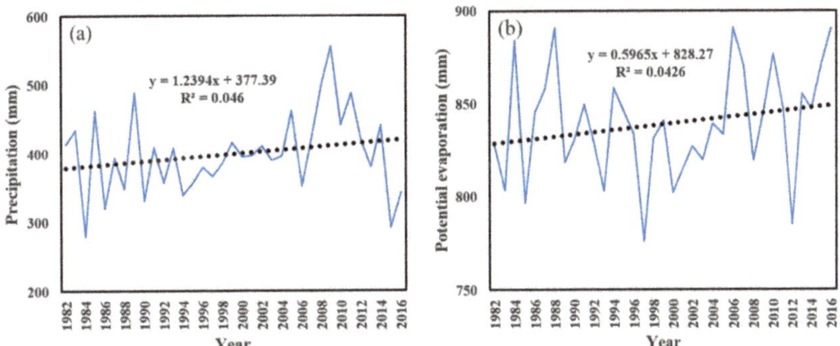

Figure 3. Variation trend of average precipitation (**a**) and average potential evaporation (**b**) in the SAYR.

4.2. Mutation Analysis of Runoff and NDVI

In this study, the Mann–Kendall method was used to analyze runoff and NDVI data in the SAYR, and the results are shown in Figures 4 and 5.

It can be found from Figure 4 that the UF and UB curve calculated from the runoff time series data in the SAYR intersected in 2004, and this intersection point was within the range of the 0.05 significance level, indicating that 2004 was the abrupt change year of runoff in the SAYR from 1982 to 2016.

It can be found from Figure 5 that UF and UB curve calculated from the NDVI time series data in the SAYR intersected in 1998, and this intersection point was within the range of the 0.05 significance level, indicating that 1998 was abrupt change year of NDVI in the SAYR from 1982 to 2016. This may be because Grain for Green was proposed by Chinese government in 1998, and the impact of human activities on vegetation change has increased since 1999.

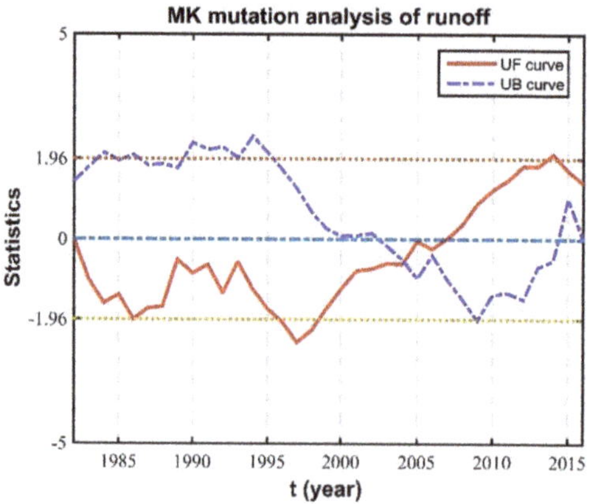

Figure 4. Mann–Kendall mutation test result of runoff in the SAYR from 1982 to 2016.

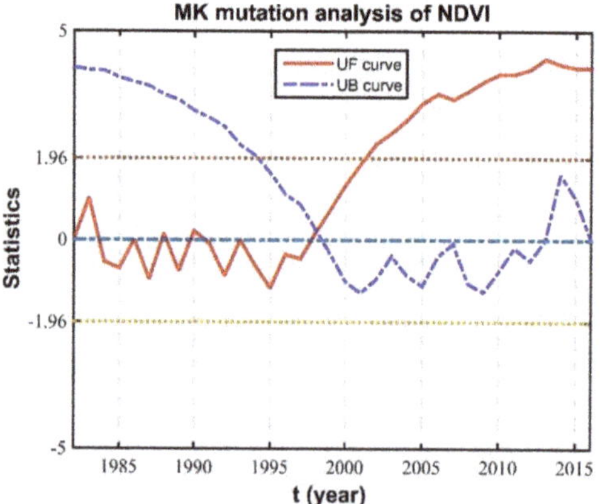

Figure 5. Mann–Kendall mutation test result of NDVI in the SAYR from 1982 to 2016.

4.3. Assessment of the Contribution Rates of Climate and Anthropic Factors to Runoff and Vegetation Changes

Identifying the abrupt change year of runoff and NDVI was the premise for dividing time series data into different periods (i.e., base period and changing period). If there was no abrupt change year of runoff and NDVI data, we did not divide time series data into different periods.

According to mutation analysis results of runoff time series data at Zhimenda hydrological station from 1982 to 2016, the Zhimenda hydrological station 1982–2016 data were divided into two periods: the base period (1982–2004) and the change period (2005–2016), and the eigenvalues of meteorological and hydrological variables for different periods in the Zhimenda hydrological station were calculated (Table 1).

Table 1. Eigenvalues of meteorological and hydrological variables in the SAYR.

Hydrological Station	Period	ET_0/mm	R/mm	P/mm	ω	R/P	ET_0/P
Zhimenda	1982–2004	831.74	90.57	386.35	1.25	0.23	2.15
	2005–2016	852.93	115.06	425.29	1.18	0.27	2.01

In order to further evaluate the temporal evolution characteristics of the impact of climate factors and underlying surface parameters on runoff, according to

Equations (11)–(13), the elastic coefficients of precipitation, potential evaporation, and underlying surface characteristic parameters in different years on runoff change of Zhimenda hydrological station were calculated (Figure 6).

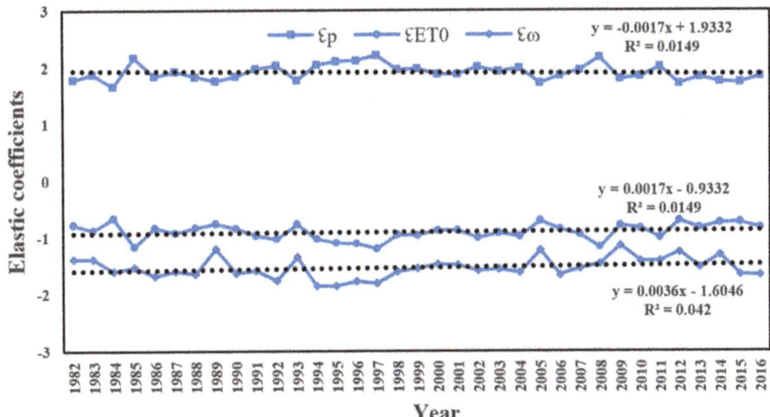

Figure 6. Change trend of precipitation elasticity, potential evapotranspiration elasticity, and underlying surface parameter elasticity of runoff in the SAYR from 1982–2016.

It can be seen from Figure 6 that the absolute values of εP, $\varepsilon ET0$, and $\varepsilon \omega$ in for Zhimenda hydrological station all showed a decreasing trend, indicating that the sensitivity of runoff change in the SAYR to climate factors and underlying surface parameters is decreasing.

According to Table 1, the differences of average annual precipitation, potential evaporation, and underlying surface parameter in the base period (1982–2004) and change period (2005–2016) were calculated and are displayed in Table 2. Then, according to Equations (15)–(17), the runoff variation amount for Zhimenda hydrological station caused by climate factors (precipitation, potential evaporation) and underlying surface condition change were calculated, as displayed in Table 2. Finally, the contribution rate of climate (precipitation, evaporation) and anthropic factors to runoff change of Zhimenda hydrological station was calculated according to Equations (18)–(21), as shown in Table 2.

Table 2. Attribution analysis of runoff variation in the SAYR.

Hydrological Station	ε_P	ε_{ET0}	ε_ω	ΔR	ΔP	ΔET_0	$\Delta\omega$	ΔR_P	ΔR_{ET0}	ΔR_L	ηR_P	ηR_{ET0}	ηR_H
Zhimenda	1.90	−0.90	−1.54	24.49	38.94	21.19	−0.07	18.34	−2.26	24.14	75.98%	−9.35%	33.37%

The contribution rates of climate and anthropic factors to runoff variation in Zhimenda hydrological station are 66.63% and 33.37%, respectively. In general, climate change plays a major role in increasing runoff, and precipitation has a more significant effect on runoff increase than the reference evapotranspiration.

This conclusion is similar to that of Liu et al. [27]. However, there are some differences in the specific contribution rate value, and this may be due to the following factors: (1) The time ranges of runoff data are different; (2) different methods are used to calculate the contribution rate of climatic (precipitation, potential evaporation) and anthropic factors to runoff variation.

According to the abrupt analysis results of NDVI data of the SAYR from 1982 to 2016, the NDVI data in the SAYR from 1982 to 2016 can be divided into two periods: the base period (1982–1998) and the change period (1999–2016). Previous studies have found that monthly NDVI is closely related to monthly precipitation and potential evaporation [43,44], and NDVI change can be used to characterize vegetation change. Referring to the method proposed by Li et al. [45], this study first established the multiple linear regression equation between monthly NDVI, monthly precipitation, and monthly potential evapotranspiration in the period of 1982–1998, and then the contribution rate of climate change and anthropic factors to vegetation change was quantitatively analyzed according to Equations (25)–(28) (Table 3). The results showed that compared to the period of 1982–1998, anthropic factors played a major role in increasing vegetation coverage in the period of 1999–2016, with a contribution rate of 61.44%, and the contribution rate of climate factors was 38.56%.

Table 3. Attribution analysis of vegetation change in the SAYR.

Fitting Equation	$\triangle NDVI$	$\overline{NDVI_{T1}}$	$\overline{NDVI_{T2,S}}$	$\overline{NDVI_{T2}}$	$\eta NDVI_C$ (%)	$\eta NDVI_H$ (%)
$NDVI = 1.7912 \times 10^{-3}P + 7.1204 \times 10^{-4}ET_0 + 0.1057$ ($R^2 = 0.68$)	0.0197	0.2122	0.2198	0.2319	38.56	61.44

5. Discussion

Although the change trend and mutation characteristics of runoff and vegetation in the SAYR were analyzed, and the impact of climate change and anthropic factors on runoff and vegetation change were calculated, there are still some uncertainties. The data of precipitation and potential evaporation in the study area are interpolated from low-density meteorological station data in and around the SAYR, but there are still some deviations between the interpolation results and their actual distribution. The change of catchment storage water for multi-year water balance is assumed as 0, and we assumed that human activities and climate change do not affect each other and are independent factors. This study also assumed that the base period is only affected by climate change. All of these factors will lead to some uncertainties in the research results.

6. Conclusions

This study revealed the change trend and mutation characteristics of runoff and vegetation in the SAYR, and the contribution rates of climate change and anthropic factors to runoff change in the SAYR were quantitatively calculated using the Budyko hypothesis and multiple linear regression method, which provides theoretical support for water resource management and ecological protection in the SAYR.

The results showed that the runoff, NDVI, precipitation, and potential evaporation in the SAYR from 1982 to 2016 all showed an increasing trend. The abrupt change years of runoff and NDVI data in the SAYR from 1982 to 2016 were 2004 and 1998, respectively. Anthropic factors play a major role in annual runoff and vegetation change, with contribution rates of 75.98% and 61.44%. In the follow-up study, we will try to quantitatively analyze the contribution rate of climate and human factors to vegetation and runoff changes in different seasons.

Author Contributions: Conceptualization, G.J. and L.W.; Data curation, G.J., H.S. and H.W.; Funding acquisition, L.W. and H.W.; Methodology, G.J. and H.S.; Project administration, L.W. and H.W.; Writing—original draft, G.J., H.S. and L.W.; Writing—review and editing, G.J., H.W. and L.W. All authors have read and agreed to the published version of the manuscript.

Funding: This research was funded by the National Natural Science Foundation of China (Grant No. 41901239), Second Tibetan Plateau Scientific Expedition and Research Program (Grant No. 2019QZKK0608), Soft Science Research of Technology Development Projects of Henan Province (Grant No. 192400410085) and China Postdoctoral Science Foundation (Grant No. 2018M640670).

Institutional Review Board Statement: Not applicable.

Informed Consent Statement: Not applicable.

Data Availability Statement: Not applicable.

Conflicts of Interest: The authors declare no conflict of interest.

Appendix A

Based on the abrupt change years of runoff in the SAYR, the runoff time series data are divided into two periods: the base period and the change period. The average annual precipitation in the base period is expressed as P_1, and the average annual precipitation in the change period is expressed as P_2. The variation of annual precipitation can be expressed as:

$$\Delta P = P2 - P1 \tag{A1}$$

Similarly, the variation of the potential evaporation ($\Delta ET0$) and characteristic parameters of underlying surface ($\Delta \omega$) can be calculated by Equations (A2) and (A3):

$$\Delta ET0 = ET0_2 - ET0_1 \tag{A2}$$

$$\Delta \omega = \omega 2 - \omega 1 \tag{A3}$$

The elasticity coefficient is the ratio of the change rate of two interrelated indexes in a certain period. The elastic coefficients of precipitation (ε_P), potential evaporation (ε_{ET0}), and underlying surface parameters (ε_ω) to runoff can be expressed by Equations (A4)–(A6), respectively.

$$\varepsilon_P = \frac{\Delta P}{P} \bigg/ \frac{\Delta R}{R} \tag{A4}$$

$$\varepsilon_{ET0} = \frac{\Delta ET0}{ET0} \bigg/ \frac{\Delta R}{R} \tag{A5}$$

$$\varepsilon_\omega = \frac{\Delta \omega}{\omega} \bigg/ \frac{\Delta R}{R} \tag{A6}$$

References

1. Highland, L.M. Geographical Overview of the Three Gorges Dam and Reservoir, China—Geologic Hazards and Environmental Impacts. U.S. Geological Survey, Commonwealth of Virginia, 2008. Available online: https://pubs.usgs.gov/of/2008/1241/ (accessed on 5 June 2021).
2. Li, Z.; Li, Z.; Song, L.; Ma, J. Characteristic and factors of stable isotope in precipitation in the source region of the Yangtze River. *Agric. For. Meteorol.* **2020**, *281*, 107825.
3. Shao, Q.; Zhao, Z.; Liu, J.; Fan, J. The characteristics of land cover and macroscopical ecology changes in the Source Region of Three Rivers in Qinghai-Tibet plateau during last 30 years. *Geogr. Res.* **2010**, *29*, 363–366. (In Chinese)
4. Wu, D.; Zhao, X.; Liang, S.; Tao, Z.; Zhao, W. Time-lag effects of global vegetation responses to climate change. *Glob. Chang. Biol.* **2015**, *21*, 3520–3531. [CrossRef]
5. Zheng, H. Birth of the Yangtze River: Age and tectonic-geomorphic implications. *Natl. Sci. Rev.* **2015**, *2*, 438–453. [CrossRef]
6. Yang, J.; Wang, Y.; Luo, G.; Xue, D.; Gang, H. Patterns and Structures of Land Use Change in the Three Rivers Headwaters Region of China. *PLoS ONE* **2015**, *10*, e0119121. [CrossRef]
7. Kusunoki, S.; Ose, T.; Hosaka, M. Emergence of unprecedented climate change in projected future precipitation. *Sci. Rep.* **2020**, *10*, 1–8. [CrossRef]
8. Niyogi, D.; Jamshidi, S.; Smith, D.; Kellner, O. Evapotranspiration Climatology of Indiana Using In Situ and Remotely Sensed Products. *J. Appl. Meteorol. Climatol.* **2020**, *59*, 2093–2111. [CrossRef]
9. Li, Z.; Huang, S.; Liu, D.; Leng, G.; Zhou, S.; Huang, Q. Assessing the effects of climate change and human activities on runoff variations from a seasonal perspective. *Stoch. Environ. Res. Risk Assess.* **2020**, *34*, 575–592. [CrossRef]
10. Jamshidi, S.; Zand-Parsa, S.; Naghdyzadegan Jahromi, M.; Niyogi, D. Application of a simple Landsat-MODIS fusion model to estimate evapotranspiration over a heterogeneous sparse vegetation region. *Remote Sens.* **2019**, *11*, 741. [CrossRef]
11. Jamshidi, S.; Zand-parsa, S.; Pakparvar, M.; Niyogi, D. Evaluation of evapotranspiration over a semiarid region using multiresolution data sources. *J. Hydrometeorol.* **2019**, *20*, 947–964. [CrossRef]
12. Zhou, Y.; Lai, C.; Wang, Z.; Chen, X.; Zeng, Z.; Chen, J.; Bai, X. Quantitative evaluation of the impact of climate change and human activity on runoff change in the Dongjiang River Basin, China. *Water* **2018**, *10*, 571. [CrossRef]
13. Tietjen, B.; Schlaepfer, D.R.; Bradford, J.B.; Lauenroth, W.K.; Hall, S.A.; Duniway, M.C.; Hochstrasser, T.; Jia, G.; Munson, S.M.; Pyke, D.A.; et al. Climate change-induced vegetation shifts lead to more ecological droughts despite projected rainfall increases in many global temperate drylands. *Glob. Chang. Biol.* **2017**, *23*, 2743–2754. [CrossRef]

14. Chen, B.; Zhang, X.; Tao, J.; Wu, J.; Wang, J.; Shi, P.; Zhang, Y.; Yu, C. The impact of climate change and anthropogenic activities on alpine grassland over the Qinghai-Tibet Plateau. *Agric. For. Meteorol.* **2014**, *189*, 11–18. [CrossRef]

15. Sun, Q.; Li, B.; Zhou, C.; Li, F.; Zhang, Z.; Ding, L.; Zhang, T.; Xu, L. A systematic review of research studies on the estimation of net primary productivity in the Three-River Headwater Region. *China J. Geogr. Sci.* **2017**, *27*, 161–182. [CrossRef]

16. Jiang, L.; Jiapaer, G.; Bao, A.; Guo, H.; Ndayisaba, F. Vegetation dynamics and responses to climate change and human activities in Central Asia. *Sci. Total Environ.* **2017**, *599*, 967–980. [CrossRef] [PubMed]

17. Yan, D.; Lai, Z.; Ji, G. Using Budyko-Type Equations for Separating the Impacts of Climate and Vegetation Change on Runoff in the Source Area of the Yellow River. *Water* **2020**, *12*, 3418. [CrossRef]

18. Ji, G.; Lai, Z.; Xia, H.; Liu, H.; Wang, Z. Future Runoff Variation and Flood Disaster Prediction of the Yellow River Basin Based on CA-Markov and SWAT. *Land* **2021**, *10*, 421. [CrossRef]

19. Liu, H.; Wang, Z.; Ji, G.; Yue, Y. Using Budyko-Type Equations for Separating the Impacts of Climate and Vegetation Change on Runoff in the Source Area of the Yellow River. *Water* **2020**, *12*, 3501. [CrossRef]

20. Chen, T.; Liang, S.; Qian, K.; Wan, L. Regularity and cause of vegetation coverage changes in the headwaters of the Changjiang River over the last 22 years. *Earth Sci. Front.* **2008**, *15*, 323–331. (In Chinese)

21. Yao, Y.; Yang, J.; Wang, R.; Lu, D.; Zhang, X. Responses of net primary productivity of natural vegetation to climatic change over source regions of Yangtze River in 1959–2008. *J. Glaciol. Geocryol.* **2011**, *33*, 1286–1293. (In Chinese)

22. Liu, L.; Cao, W.; Shao, Q. Change of ecological condition in the headwater of the Yangtze River before and after the implementation of the ecological conservation and construction project. *J. Geo-Inf. Sci.* **2016**, *18*, 1069–1076. (In Chinese)

23. Qian, K.; Wang, X.; Lv, J.; Li, W. The wavelet correlative analysis of climatic impacts on runoff in the source region of Yangtze River, in China. *Int. J. Climatol.* **2014**, *34*, 2019–2032. [CrossRef]

24. Li, L.; Shen, H.; Dai, S.; Li, H.; Xiao, J. Response of water resources to climate change and its future trend in the source region of the Yangtze River. *J. Geogr. Sci.* **2013**, *23*, 208–218. [CrossRef]

25. Luo, Y.; Qin, N.; Zhou, B.; Li, J.; Liu, J.; Wang, C.; Pang, Y. Changes of runoff in the source region of the Yangtze River from 1961 to 2016. *Soil Water Conserv. Res.* **2019**, *26*, 123–128. (In Chinese)

26. Xu, X.; Yan, J.; Liang, X. Characteristics of runoff variation and degree of human influence in the Three-River Headwaters Region. *Arid Area Study* **2009**, *26*, 88–93. (In Chinese)

27. Liu, J.; Chen, J.; Xu, J.; Lin, Y.; Yuan, Z.; Zhou, M. Attribution of Runoff Variation in the Headwaters of the Yangtze River Based on the Budyko Hypothesis. *Int. J. Environ. Res. Public Health* **2019**, *16*, 2506. [CrossRef] [PubMed]

28. Li, H.; Liu, G.; Fu, B. Response of vegetation to climate change and human activity based on NDVI in theThree-River Headwaters region. *Acta Ecol. Sin.* **2011**, *31*, 5495–5504. (In Chinese)

29. Li, Y.; Li, Q.; Liu, X.; Duan, S.; Cai, Y. Analysis of Runoff Variation in Source Region of Yangtze River. *J. China Hydrol.* **2017**, *37*, 92–95. (In Chinese)

30. Su, Y.; Chen, X.; Li, Y.; Liao, J.; Ye, Y.; Zhang, H.; Huang, N.; Kuang, Y. China's 19-year city-level carbon emissions of energy consumptions, driving forces and regionalized mitigation guidelines. *Renew. Sustain. Energy Rev.* **2014**, *35*, 231–243. [CrossRef]

31. Wang, H.; Ji, G.; Xia, J. Analysis of Regional Differences in Energy-Related PM2.5 Emissions in China: Influencing Factors and Mitigation Countermeasures. *Sustainability* **2019**, *11*, 1409. [CrossRef]

32. Xia, H.; Wang, H.; Ji, G. Regional Inequality and Influencing Factors of Primary PM Emissions in the Yangtze River Delta, China. *Sustainability* **2019**, *11*, 2269. [CrossRef]

33. Zeng, Z.; Liu, J.; Savenije, H.H. A simple approach to assess water scarcity integrating water quantity and quality. *Ecol. Indic.* **2013**, *34*, 441–449. [CrossRef]

34. Yan, X.; Bao, Z.; Zhang, J.; Wang, G.; He, R.; Liu, C. Quantifying contributions of climate change and local human activities to runoff decline in the upper reaches of the Luanhe River basin. *J. Hydro-Environ. Res.* **2020**, *28*, 67–74. [CrossRef]

35. Caracciolo, D.; Pumo, D.; Viola, F. Budyko's based method for annual runoff characterization across different climatic areas: An application to United States. *Water Resour. Manag.* **2018**, *32*, 3189–3202. [CrossRef]

36. Ji, G.; Wu, L.; Wang, L.; Yan, D.; Lai, Z. Attribution Analysis of Seasonal Runoff in the Source Region of the Yellow River Using Seasonal Budyko Hypothesis. *Land* **2021**, *10*, 542. [CrossRef]

37. Zhang, X.; Dong, Q.; Cheng, L.; Xia, J. A Budyko-based framework for quantifying the impacts of aridity index and other factors on annual runoff. *J. Hydrol.* **2019**, *579*, 124224. [CrossRef]

38. Wu, J.; Miao, C.; Wang, Y.; Duan, Q.; Zhang, X. Contribution analysis of the long-term changes in seasonal runoff on the Loess Plateau, China, using eight Budyko-based methods. *J. Hydrol.* **2017**, *545*, 263–275. [CrossRef]

39. Zhang, D.; Cong, Z.; Ni, G.; Yang, D.; Hu, S. Effects of snow ratio on annual runoff within the Budyko framework. *Hydrol. Earth Syst. Sci.* **2015**, *19*, 1977. [CrossRef]

40. Choudhury, B.J. Evaluation of an empirical equation for annual evaporation using field observations and results from a biophysical model. *J. Hydrol.* **1999**, *216*, 99–110. [CrossRef]

41. Yang, H.; Yang, D.; Lei, Z.; Sun, F. New analytical derivation of the mean annual water-energy balance equation. *Water Resour. Res.* **2008**, *44*, W03410. [CrossRef]

42. Xu, X.; Yang, D.; Yang, H.; Lei, H. Attribution analysis based on the Budyko hypothesis for detecting the dominant cause of runoff decline in Haihe basin. *J. Hydrol.* **2014**, *510*, 530–540. [CrossRef]

43. Gao, J.; Jiao, K.; Wu, S. Investigating the spatially heterogeneous relationships between climate factors and NDVI in China during 1982 to 2013. *J. Geogr. Sci.* **2019**, *29*, 1597–1609. [CrossRef]
44. Guo, L.; Zuo, L.; Gao, J.; Jiang, Y.; Zhang, Y.; Ma, S.; Zou, Y.; Wu, S. Revealing the Fingerprint of Climate Change in Interannual NDVI Variability among Biomes in Inner Mongolia, China. *Remote Sens.* **2020**, *12*, 1332. [CrossRef]
45. Li, J.; Peng, S.; Li, Z. Detecting and attributing vegetation changes on China's Loess Plateau. *Agric. For. Meteorol.* **2017**, *247*, 260–270. [CrossRef]

Article

Quantitatively Calculating the Contribution of Vegetation Variation to Runoff in the Middle Reaches of Yellow River Using an Adjusted Budyko Formula

Guangxing Ji [1], Junchang Huang [1], Yulong Guo [1] and Dan Yan [2,3,*]

[1] College of Resources and Environmental Sciences, Henan Agricultural University, Zhengzhou 450002, China; guangxingji@henau.edu.cn (G.J.); huangjc1014@henau.edu.cn (J.H.); gyl.zh@henau.edu.cn (Y.G.)

[2] Center for Energy, Environment & Economy Research, Zhengzhou University, No. 100 Science Avenue, Gaoxin District, Zhengzhou 450001, China

[3] Tourism Management School, Zhengzhou University, No. 100 Science Avenue, Gaoxin District, Zhengzhou 450001, China

* Correspondence: yandan@zzu.edu.cn

Abstract: The middle reaches of the Yellow River (MRYR) are a key area for carrying out China's vegetation restoration project. However, the impact of vegetation variation on runoff in the MRYR is still unclear. For quantitatively evaluating the contribution rate of vegetation variation to runoff in the MRYR, this paper quantified the relationship between Normalized Difference Vegetation Index (NDVI) and Budyko parameters (w). Then, we used multiple linear regression to quantitatively calculate the contribution rate of different factors on vegetation variation. Finally, an adjusted Budyko formula was constructed to quantitatively calculate the influence of vegetation variation on runoff. The results showed that there is a linear relationship between NDVI and Budyko parameters (w) ($p < 0.05$); the fitting parameter and constant term were 12.327 and -0.992, respectively. Vegetation change accounted for 33.37% in the MRYR. The contribution of climatic and non-climatic factors on vegetation change is about 1:99. The contribution of precipitation, potential evaporation, anthropogenic activities on the runoff variation in the MRYR are 23.07%, 13.85% and 29.71%, respectively.

Keywords: vegetation variation; runoff variation; Budyko hypothesis; attribution analysis

Citation: Ji, G.; Huang, J.; Guo, Y.; Yan, D. Quantitatively Calculating the Contribution of Vegetation Variation to Runoff in the Middle Reaches of Yellow River Using an Adjusted Budyko Formula. *Land* **2022**, *11*, 535. https://doi.org/ 10.3390/land11040535

Academic Editors: Wiktor Halecki, Dawid Bedla, Marek Ryczek and Artur Radecki-Pawlik

Received: 13 March 2022
Accepted: 5 April 2022
Published: 7 April 2022

Publisher's Note: MDPI stays neutral with regard to jurisdictional claims in published maps and institutional affiliations.

Copyright: © 2022 by the authors. Licensee MDPI, Basel, Switzerland. This article is an open access article distributed under the terms and conditions of the Creative Commons Attribution (CC BY) license (https:// creativecommons.org/licenses/by/ 4.0/).

1. Introduction

In the past few decades, many ecological protection projects carried out in China have led to a significant increase in vegetation [1,2]. The middle reaches of the Yellow River (MRYR) are the key area for the implementation of the vegetation restoration project in China [3,4]. Vegetation index is a simple, effective and empirical measure of surface vegetation, so the Normalized Difference Vegetation Index (NDVI) was used to characterize the vegetation change in the MRYR. Some studies show that NDVI in the MRYR has increased significantly, and its vegetation coverage has been restored [5–7].

Vegetation change can significantly change runoff by changing hydrological processes, such as vegetation transpiration and interception evaporation, and then affect the available amount of watershed runoff [8–10]. The debate on the relationship between "vegetation and water" dates back to at least the mid-19th century [11]. Although most studies show that the increase in vegetation will lead to a decrease in water yield (called negative impact here), only a few studies show that the increase in vegetation has little impact on water yield [12,13] (called no impact here) and even leads to an increase in water yield [14,15] (called positive impact here). Interestingly, the areas with positive and no impact are mostly in humid areas and large watersheds with complex terrain. In the past few decades, the spatial variability and periodicity of precipitation and runoff in the MRYR have changed [16–19]. Therefore, quantitatively calculating the influence of vegetation variation on runoff variation is helpful to deeply understand the response mechanism of the water cycle process to the underlying surface condition changes, and has theoretical and practical significance for improving watershed water resources management measures. It is useful for the utilization and development of water resources in the Yellow River, the protection of the human living environment and the sustainable development of the economy.

Many scholars have carried out studies for quantitatively calculating the influence of vegetation variation on runoff variation in the MRYR [20–22]. The runoff decreased with the improvement of vegetation coverage, seen by analyzing the statistical relationship between vegetation coverage and runoff in most basins in the MRYR [23]. Liang et al. [24] quantitatively analyzed the influence of ecological restoration measures for the streamflow variation of most watersheds on the Loess Plateau from 1961 to 2009, based on Budyko theory, and found that ecological restoration measures are the leading factor for the decline in runoff in most watersheds. Li et al. [25] found that the Normalized Difference Vegetation Index (NDVI) in most basins in the Loess Plateau increased significantly after 2000, while the runoff decreased significantly. A higher evapotranspiration rate from restored vegetation is the primary reason for the reduced runoff coefficient. Zhang et al. [26] quantitatively evaluated the influence of vegetation cover variation on streamflow variation in the MRYR over the past 37 years. It was shown that vegetation variation is the leading driving force of streamflow decline from 2000 to 2010. Wang et al. [27] discovered that the vegetation coverage in the MRYR increased by 29.72% from 1982 to 2018. The average runoff at each station showed a downward trend from 1961 to 2018, and ecological restoration played an important role in the decline in runoff in the MRYR. However, the contribution of vegetation variation on streamflow variation in the MRYR has not been calculated quantitatively.

Therefore, in order to quantitatively evaluate the contribution rate of vegetation variation on streamflow variation in the middle reaches of the Yellow River (MRYR), this paper analyzes its hydro-meteorological characteristics from 1982 to 2015, identifies the mutation year of streamflow, and quantifies the relationship between the NDVI and a Budyko parameter (w). Then, multiple linear regression is adopted to distinguish the contribution rate of different factors on vegetation variation. Finally, an adjusted Budyko formula is constructed to quantitatively calculate the influence of vegetation variation on runoff in the MRYR. This work is useful to understand the influence of vegetation variation on the evolution of water resources in the MRYR, and also supply important insights for water resources management.

2. Research Region and Data

2.1. Research Region

The middle reaches of the Yellow River (MRYR) are in the area of the Toudaoguai-Huayuankou hydrological station (Figure 1). The length of this reach is about 1234.6 km, its area accounts for about 45.7% and its runoff accounts for about 44.3% of the Basin. Guanzhong Plain and Hetao Plain are in the MRYR, which produce a large number of agricultural products every year. The wheat output ranks first, soybean output ranks second, and cotton output ranks fourth in China. The decreasing runoff in the MRYR has brought great harm to agricultural production and food security.

2.2. Data

(1) The annual runoff observation data of Toudaoguai and Huayuankou hydrometric stations from 1982 to 2015 were obtained from the hydrological statistical yearbook of China. In this study, the annual runoff data in the MRYR was equal to a value, which was the annual measured runoff of Huayuankou hydrometric station minus the annual measured runoff data of Toudaoguai hydrological station.

(2) 82 meteorological stations data (1982–2015) were downloaded from China Meteorological Administration. We adopted the Penman–Monteith formula for calculating the potential evaporation of 82 meteorological stations, and then the annual precipita-

tion and potential evaporation were interpolated by Kriging. Finally, the average annual precipitation and potential evaporation in the MRYR can be obtained.

$$ET_0 = \frac{0.408\Delta(R_n - G) + \gamma\frac{900}{T+273}U_2(e_a - e_d)}{\Delta + \gamma(1 + 0.34U_2)} \tag{1}$$

(3) The NDVI dataset (8×8 km^2) was sourced from the National Aeronautics and Space Administration. The time span of the dataset is from July 1981 to December 2015, its time resolution is 15 d, and its downloaded files are in NC4 format. The annual NDVI are calculated by the average value composites method, and is equal to the arithmetic mean of all grids in the study area. This method further eliminates some interference of cloud, atmosphere, solar altitude, and sensor sensitivity. Finally, the annual NDVI from 1982 to 2015 can be obtained.

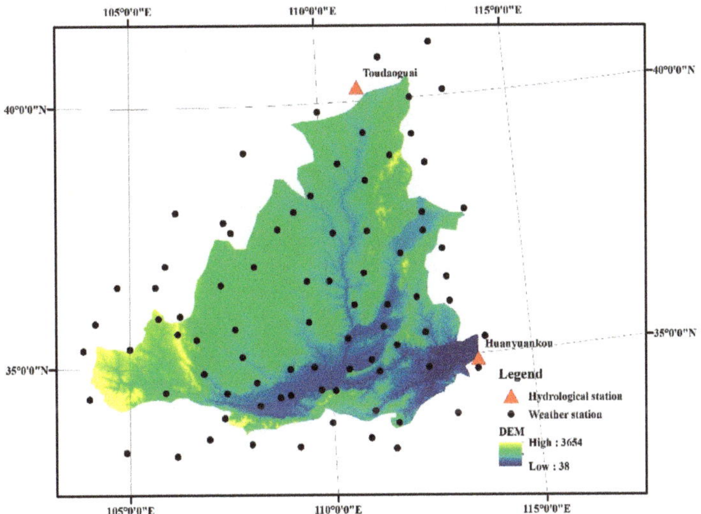

Figure 1. Location of the study area and distribution of Hydro-meteorological stations.

3. Methods

3.1. Trend Analysis Method (Slope)

The formula of *Slope* method is as follows [28,29]:

$$Slope = \frac{n\sum_{i=1}^{n} iX_i - \sum_{i=1}^{n} i\sum_{i=1}^{n} X_i}{n\sum_{i=1}^{n} i^2 - \left(\sum_{i=1}^{n} i\right)^2} \tag{2}$$

If *Slope* is greater than 0, the variable shows ever-growing properties. If *Slope* < 0, it implies that the variable gets less with the day; n means the number of years; X indicates the annual precipitation, and potential evaporation runoff.

3.2. Mann-Kendall (MK) Mutation Detection Algorithm

For X containing n data, we first construct a rank sequence. Then, we calculated S_k.

$$S_k = \sum_{i=1}^{k} r_i \quad (k = 2, 3, \ldots, i) \tag{3}$$

$$r_i = \begin{cases} +1 & x_i > x_j \\ 0 & x_i \leq x_j \end{cases} \quad (j = 2, 3, \ldots, i) \tag{4}$$

$E(S_k)$ represents the average value of S_k, $Var(S_k)$ represents the variance of S_k. Finally, $UF_1 = 0$, the Uf_k statistics can be calculated.

$$E(S_k) = \frac{n(n+1)}{4} \quad (2 \leq k \leq n) \tag{5}$$

$$Var(S_k) = \frac{n(n-1)(2n+5)}{72} \quad (2 \leq k \leq n) \tag{6}$$

$$UF_k = \frac{[S_k - E(S_k)]}{\sqrt{Var(S_k)}} \quad (k = 1, 2, \ldots, i) \tag{7}$$

Constructing the reverse order of X, and making $UB_k = UF_k$ ($k = n, n-1, \ldots, 1$), $UB_1 = 0$, we calculated the UB_k statistics in this way.

3.3. Budyko Hypothesis

There are two hypotheses in the attribution analysis of runoff change using Budyko hypothesis: (1) assuming that human activities, climate factors and vegetation do not affect each other, and are independent factors; (2) assuming that the base period is only affected by climate factors.

The multi-year scale water balance leads to Formula (8).

$$Q = P - E \tag{8}$$

where, Q represents runoff depth; P represents precipitation; E represents actual evaporation.

According to the Buduko hypothesis, there is a nonlinear function relation between E/P and dryness ratio (phi = E_0/P) of a specific watershed.

$$\frac{E}{P} = F\left(\frac{E_0}{P}\right) = F(\varphi) \tag{9}$$

E_0 is the potential evaporation (mm). The Fu-type Budyko equation is [30]:

$$F(\varphi) = 1 + \varphi + (1 + \varphi^w)^{1/w} \tag{10}$$

w represents the variable representing the underlying surface information of the basin, and is used to characterize human activities. Human activities can affect the runoff change of the Yellow River Basin through many aspects, including vegetation change, water conservancy project construction, domestic water for urban residents and agricultural irrigation water, etc. The larger the value of w, the less precipitation is transformed into runoff.

$$Q = P\left((1 + \varphi^w)^{1/w} - \varphi\right) \tag{11}$$

Li et al. [31] found that there is a good linear relationship between $NDVI$ and Budyko parameter (w) in the 26 major global river basins, which makes it possible to quantitatively evaluate the contribution rate of vegetation variation on runoff change.

$$w = a \times NDVI + b \tag{12}$$

a and b are constants for a specific watershed and can be calculated by the least squares method.

$$Q = P\left(\left(1 + \varphi^{a \times NDVI + b}\right)^{1/(a \times NDVI + b)} - \varphi\right) \tag{13}$$

The elasticity coefficient is used to measure the relative change of another variable caused by the change of one variable. ε_P, ε_{E0}, ε_w and ε_{NDVI} represent the elastic coefficient of P, E_0, w and NDVI on runoff, respectively.

$$\varepsilon_P = 1 + \frac{\varphi F'(\varphi)}{1 - F(\varphi)} \tag{14}$$

$$\varepsilon_{Eo} = -\frac{\varphi F'(\varphi)}{1 - F(\varphi)} \tag{15}$$

$$\varepsilon_w = -\frac{w F'(w)}{1 - F(\varphi)} \tag{16}$$

$$\varepsilon_{NDVI} = \varepsilon_w \frac{a \times NDVI}{a \times NDVI + b} \tag{17}$$

$$F'(\varphi) = 1 - \varphi^{w-1}(1 + \varphi^w)^{1/w-1} \tag{18}$$

$$F'(w) = \frac{(1 + \varphi^w)^{1/w} \ln(1 + \varphi^w)}{w^2} - \frac{\varphi^w (1 + \varphi^w)^{1/w-1} \ln \varphi}{w} \tag{19}$$

ΔR_P, ΔR_{E0}, ΔR_w, ΔR_{NDVI} and ΔR_H represent the value of runoff change caused by precipitation, potential evapotranspiration, underlying surface change, vegetation change and human activities.

$$\Delta R_P = \varepsilon_P \frac{R}{P} \times \Delta P \tag{20}$$

$$\Delta R_{E0} = \varepsilon_{E0} \frac{R}{E_0} \times \Delta E_0 \tag{21}$$

$$\Delta R_w = \varepsilon_w \frac{R}{w} \times \Delta w \tag{22}$$

$$\Delta R_{NDVI} = \varepsilon_{NDVI} \frac{R}{NDVI} \times \Delta NDVI \tag{23}$$

$$\Delta R_H = \Delta R_w - \Delta R_{NDVI} \tag{24}$$

ΔP, ΔE_0, Δw and $\Delta NDVI$ represent the changes of multi-year average P, E_0, w and $NDVI$ from T_1 to T_2 period.

$\Delta NDVI$ is attributed to $NDVI$ change value caused by climate factors ($\Delta NDVI_c$) and human activities ($\Delta NDVI_H$):

$$\Delta NDVI = NDVI_{T2} - NDVI_{T1} = \Delta NDVI_c + \Delta NDVI_H \tag{25}$$

Average $NDVI$ is significantly correlated with P and E_0 [32–34], so we calculated the multiple linear regression equation between $NDVI$, P and E_0 in T_1 period. There are two assumptions in the attribution analysis of vegetation change using multiple linear regression method: (1) Human activities and climate factors are independent factors. (2) The base period is only affected by climate factors. Therefore, except for climate factors, other factors affecting vegetation change are classified as human activities.

$$NDVI_{T1} = a \times P_{T1} + b \times E_{0T1} + c \tag{26}$$

$$NDVI_{T2,s} = a \times P_{T2} + b \times E_{0T2} + c \tag{27}$$

$$\Delta NDVI_H = NDVI_{T2} - NDVI_{T2,S} \tag{28}$$

$$\Delta NDVI_c = NDVI_{T2,S} - NDVI_{T1} \tag{29}$$

$$\eta_{NDVI_H} = \Delta NDVI_H / \Delta NDVI \tag{30}$$

$$\eta_{NDVI_c} = \Delta NDVI_c / \Delta NDVI \tag{31}$$

where, $NDVI_{T2,S}$ represents average $NDVI$ value under climate change in the T_2 period. $NDVI_{T1}$ and $NDVI_{T2}$ represent average $NDVI$ value of the two periods, respectively. η_{NDVI_H} and η_{NDVI_c} represent the influence of anthropic and climatic factors on vegetation variation.

$$\Delta R = \Delta R_P + \Delta R_{E0} + \eta_{NDVI_c} \times \Delta R_{NDVI} + \eta_{NDVI_H} \times \Delta R_{NDVI} + \Delta R_H \tag{32}$$

$$\eta_{Rp} = \Delta R_P / \Delta R \times 100\% \tag{33}$$

$$\eta_{R_{E0}} = \Delta R_{E0} / \Delta R \times 100\% \tag{34}$$

$$\eta_{R_{NDVIC}} = \eta_{NDVI_C} \times \Delta R_{NDVI} / \Delta R \times 100\% \tag{35}$$

$$\eta_{R_{NDVIH}} = \eta_{NDVI_H} \times \Delta R_{NDVI} / \Delta R \times 100\% \tag{36}$$

$$\eta_{R_H} = \Delta R_H / \Delta R \times 100\% \tag{37}$$

η_{Rp}, $\eta_{R_{E0}}$, $\eta_{R_{NDVIC}}$, $\eta_{R_{NDVIH}}$ and η_{R_H} represent the contribution rates of precipitation, potential evaporation, climatic vegetation change, anthropogenic vegetation change and anthropic factors on runoff.

4. Results and Analysis

4.1. Trend Analysis of Meteorological and Hydrological Elements

The changes in runoff, precipitation and potential evaporation in the MRYR from 1982 to 2015 are displayed in Figure 2. From Figure 2a, the annual runoff in the MRYR showed a significant decreasing tendency in the period of 1982–2015 ($p < 0.05$), with a slope of -3.3143×10^8 m^3/a. From Figure 2b, the average annual precipitation in the MRYR showed a tendency to increase from 1982 to 2015 ($p > 0.05$), with a slope of 0.0582 mm/a. The average annual potential evaporation in the middle reaches of the Yellow River displayed a fluctuating growth tendency in the period of 1982–2015 ($p < 0.05$), and the increase rate was 1.6357 mm/a.

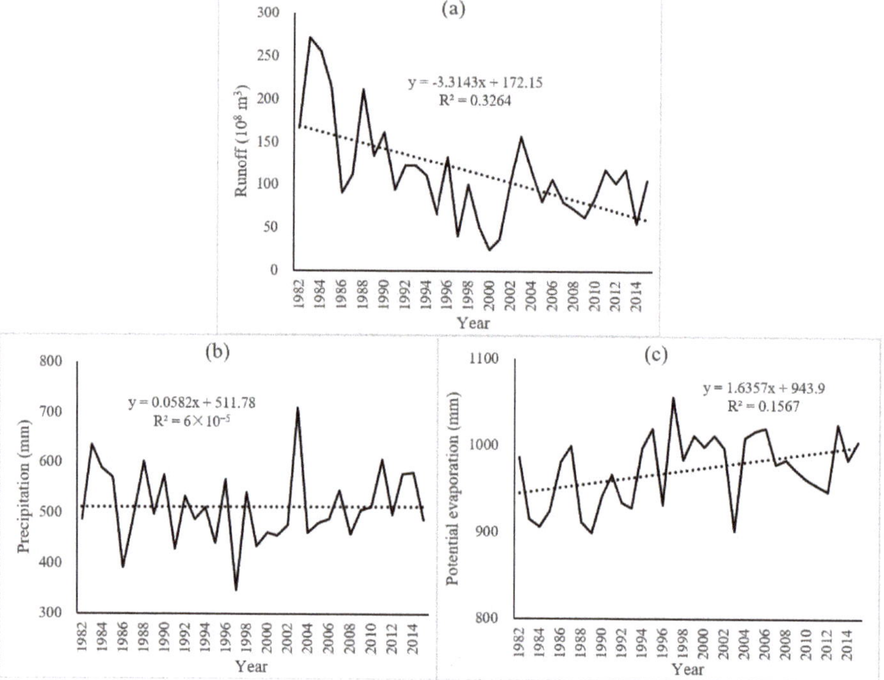

Figure 2. Changes in annual runoff (**a**), precipitation (**b**) and potential evaporation (**c**) in MRYR.

4.2. Mutation Year Identification

Because the Mann-Kendall mutation test algorithm is not disturbed by a few outliers and has the advantages of simple calculation, it is widely used to identify the mutation points of hydro-meteorological elements [35–38].

In the Mann-Kendall mutation detection algorithm, UF statistics are a sequence of statistics calculated by the order of time series X, and UB statistics are a sequence of statistics calculated by the inverse order of time series X. In practical application, it is generally considered that the result is credible at the significance level of 0.05. If the UF statistics line and UB statistics line have intersections and these intersections are within the range of 0.05, this is significant. The year corresponding to the intersection is the mutation year of the time series data.

UF statistics of runoff time series data in the MRYR intersected with UB statistics in 1989 (Figure 3), and this intersection is within the range of 0.05 (significant level), indicating that 1989 is a runoff mutation year in the MRYR.

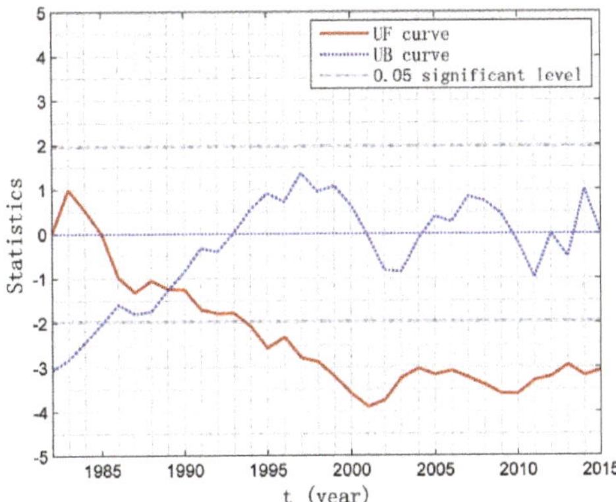

Figure 3. Mann-Kendall catastrophe analysis result of runoff in MRYR from 1982 to 2015.

UF statistics of precipitation time series data in the MRYR intersected with UB statistics in 1984, 2010, 2012 (Figure 4), and these intersections are all within the range of the 0.05 significant level, indicating that 1984, 2010, 2012 are precipitation mutation years in the MRYR.

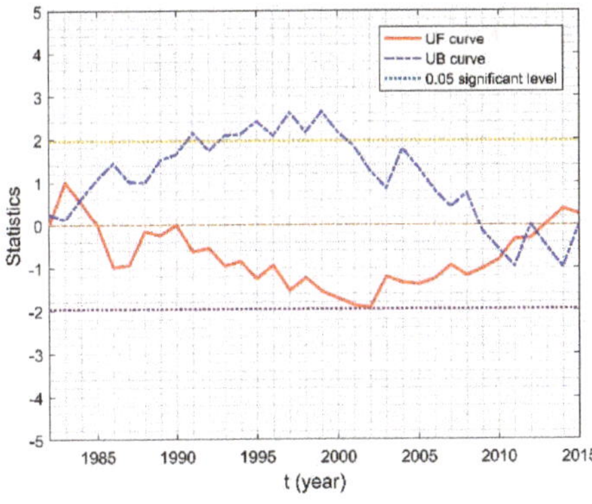

Figure 4. Mann-Kendall catastrophe analysis result of precipitation in MRYR from 1982 to 2015.

4.3. The Relationship between NDVI and w

In Budyko's hypothesis, the parameter (w) reflects the change in the underlying surface, so the parameter (w) should change on an annual or multi-year scale. For the Fu-type Budyko equation, if the elements (Q, P, E_0) in the equation can be determined, the parameter (w) can be solved. Using the hydrological and meteorological data in the MRYR from 1982 to 2015, the parameters (w) from 1982 to 2015 are obtained. Figure 5 displayed the variation tendency of NDVI and underlying surface parameters (w) in the MRYR from 1982 to 2015. The NDVI and underlying surface parameters (w) in the middle reaches of the Yellow River increased significantly ($p < 0.05$), and the slopes were 0.017/10a and 0.268/10a, respectively.

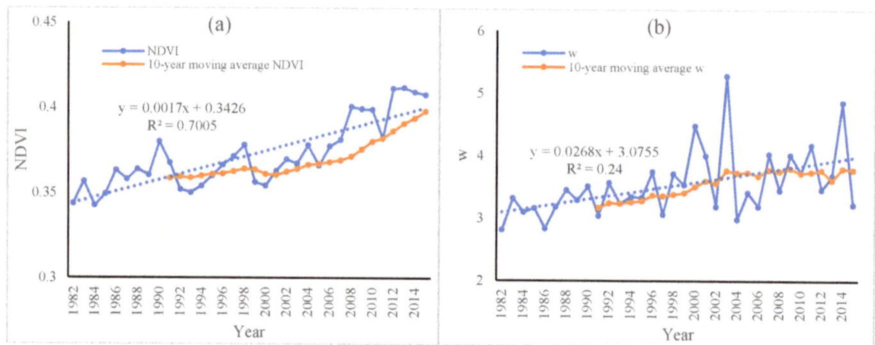

Figure 5. Variation trend of NDVI (**a**) and underlying surface parameters (w) (**b**) in the middle reaches of the Yellow River from 1982 to 2015.

In this paper, the values of underlying surface parameters (w) and NDVI are processed by a 10-year moving average, and then the regression coefficients (a and b) are obtained by least square fitting, a = 12.327, b = −0.992. The R^2 of equation fitting is 0.4465, and is significant ($p < 0.05$), indicating that the fitting result is good (Figure 6).

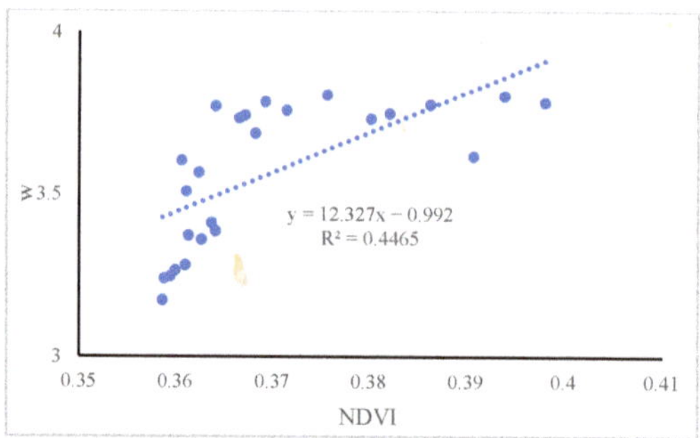

Figure 6. The relationship between NDVI and underlying surface parameter (w).

4.4. Quantifying the Influence of Vegetation Variation on Runoff

Because we want to quantitatively calculate the contribution rate of different factors on runoff change (Figure 7), 1982–2015 is separated into T_1 (1982–1989) and T_2 (1990–2015), according to the catastrophe analysis results of runoff data in the MRYR. Firstly, we analyzed the correlation between *NDVI*, *P* and E_0 (Figure 8), implying that NDVI is significantly correlated with *P* and E_0. Next, we calculated the linear function relation between *NDVI*, *P* and E_0 in T_1 period. Finally, we analyzed the contribution rate of different factors on vegetation change, using formulas 27–31 (Table 1). The results showed that the contribution of climatic and non-climatic factors on vegetation variation is close to 1:99.

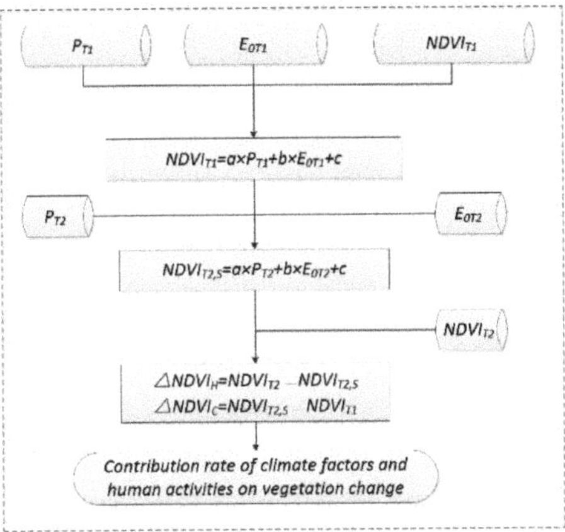

Figure 7. A flow chart for calculating the contribution of different factors on vegetation change.

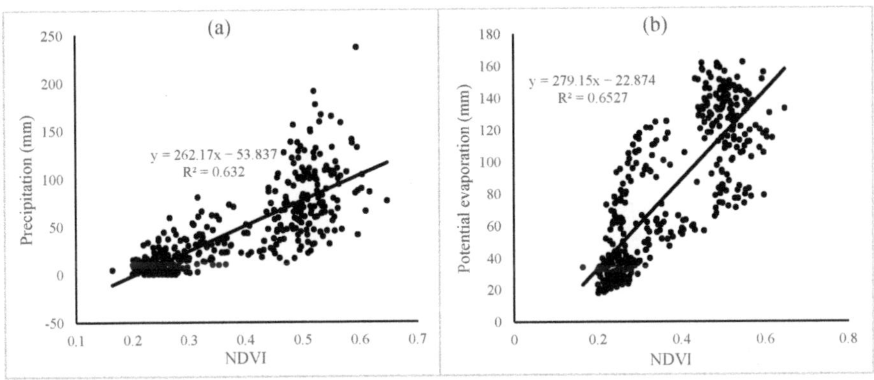

Figure 8. The relationship between NDVI and precipitation (**a**) and potential evaporation (**b**).

Table 1. Attribution analysis of vegetation variation in MRYR.

T_1	T_2	Fitting Equation	$\Delta NDVI$	$NDVI_{T1}$	$NDVI_{T2,S}$	$NDVI_{T2}$	η_{NDVI_C}	η_{NDVI_H}
1982–1989	1990–2015	$NDVI = 1.495 \times 10^{-3}P + 1.344 \times 10^{-3}ET_0 + 0.1807$ $(R^2 = 0.84)$	0.0227	0.3548	0.3551	0.3775	0.97%	99.03%

Shown in Table 2, compared with the T_1 period, the potential evaporation in the T_2 period increased by 41.70 mm, precipitation and runoff depth decreased by 25.89 mm and 24.49 mm, respectively, underlying surface parameters (w) and NDVI increased by 0.52 and 0.023, respectively.

Table 2. The value of climate, hydrology and NDVI variables in different periods in MRYR.

Periods	E_0/mm	P/mm	R/mm	w	NDVI
T (1982–2015)	972.53	512.80	31.52	3.55	0.373
T_1 (1982–1989)	940.64	532.60	50.25	3.15	0.355
T_2 (1990–2015)	982.34	506.71	25.76	3.67	0.378
T_2–T_1	41.70	−25.89	−24.49	0.52	0.023

Table 3 displays the influence of climatic and anthropogenic factors on the annual runoff of the post-change period. Contrasted to T_1, runoff variation amount caused by precipitation, potential evapotranspiration, vegetation variation and anthropogenic activities in the T_2 is -5.42 mm, -3.25 mm, -7.84 mm, and -6.98 mm. Their contributions were 23.07%, 13.85%, 33.37%, and 29.71%, respectively.

Table 3. Attribution analysis of runoff variation in MRYR.

ε_P	ε_{E0}	ε_ω	ε_{NDVI}	ΔR_P	ΔR_{E0}	ΔR_{NDVI}	ΔR_{hum}	ηR_P	ηR_{E0}	ηR_{NDVIC}	ηR_{NDVIH}	ηR_{hum}
3.41	-2.41	-3.20	-4.08	-5.42	-3.25	-7.84	-6.98	23.07%	13.85%	0.32%	33.05%	29.71%

5. Conclusions and Discussion

Using NDVI, meteorological data and measured runoff data, based on an adjusted Fu-type Budyko equation, this paper quantitatively analyzed the contribution of vegetation variation on runoff in the MRYR from 1982 to 2015. The results showed that the influence of climatic and anthropogenic factors on vegetation variation is about 1:99. Vegetation change played a major role on runoff in the MRYR. Precipitation, potential evaporation and human activities account for 23.07%, 13.85% and 29.71% in runoff variation, respectively.

Although this paper carries out strict quality control on the data and model, some non-determinacy remains. Firstly, precipitation and potential evaporation data are interpolated from the data of meteorological stations in and around the MRYR, which will bring some non-determinacy. Secondly, the precondition of the study is that all variables are independent and do not affect each other. For example, when calculating the partial derivative of multi-year evapotranspiration to runoff, it is required that the factors, such as vegetation change, precipitation and human activities, are independent. However, in practice, the underlying surface and climate system are a whole. These factors interact and affect each other, which will have some non-determinacy on the research results.

This paper quantitatively calculated the influence of vegetation variation on runoff in the MRYR, but in order to reveal the influence mechanism of vegetation on water cycle, it is necessary to analyze how vegetation affects the hydrological process of "precipitation-interception-infiltration-evapotranspiration-runoff". In the follow-up study, we will combine the distributed hydrological model HLMS with the PML model, coupling leaf area index information to build a distributed Hydrothermal Coupling Model [39–41], for clarifying the response of water heat balance to vegetation variation. In addition, in the process of quantitatively calculating the contribution rate of different factors on runoff change, in addition to climate factors and vegetation change, water conservancy project construction, urban residents' domestic water and agricultural irrigation water are classified as human activity factors. We will quantitatively calculate the contribution rate of urban water, industrial water, agricultural water and water conservancy project construction on runoff change in the follow-up study.

Author Contributions: Conceptualization: G.J. and D.Y.; Data curation: G.J. and J.H.; Methodology: G.J., Y.G. and D.Y.; Writing—original draft: G.J. and D.Y.; Writing—review and editing: G.J., J.H., Y.G. and D.Y. All authors have read and agreed to the published version of the manuscript.

Funding: This research was funded by the National Key R&D Program of China (2021YFD1700900), Soft Science Research of Technology Development Projects of Henan Province (212400410077), Think tanks key research projects of Colleges and Universities of Henan Province (2022ZKYJ07), Young backbone teachers program of Henan Province (2019GGJS048) and the special fund for top talents in Henan Agricultural University (30501031).

Institutional Review Board Statement: Not applicable.

Informed Consent Statement: Not applicable.

Data Availability Statement: Not applicable.

Conflicts of Interest: We declare no conflict of interest.

References

1. Wei, X.; Sun, G.; Liu, S.; Jiang, H.; Zhou, G.; Dai, L. The forest-streamflow relationship in China: A 40-year retrospect. *J. Am. Water Resour. Assoc.* **2008**, *44*, 1076–1085. [CrossRef]
2. Lü, Y.; Zhang, L.; Feng, X.; Zeng, Y.; Fu, B.; Yao, X.; Li, J.; Wu, B. Recent ecological transitions in China: Greening, browning, and influential factors. *Sci. Rep.* **2015**, *5*, 8732. [CrossRef]
3. Piao, S.; Yin, G.; Tan, J.; Cheng, L.; Huang, M.; Li, Y.; Liu, R.; Mao, J.; Myneni, R.B.; Peng, S. Detection and attribution of vegetation greening trend in China over the last 30 years. *Glob. Chang. Biol.* **2015**, *21*, 1601–1609. [CrossRef] [PubMed]
4. Feng, X.; Fu, B.; Piao, S.; Wang, S.; Ciais, P.; Zeng, Z.; Lü, Y.; Zeng, Y.; Li, Y.; Jiang, X. Revegetation in China's Loess Plateau is approaching sustainable water resource limits. *Nat. Clim. Chang.* **2016**, *6*, 1019–1022. [CrossRef]
5. Liu, J.; Li, Z.; Zhang, X.; Li, R.; Liu, X.; Zhang, H. Responses of vegetation cover to the Grain for Green Program and their driving forces in the He-Long region of the middle reaches of the Yellow River. *J. Arid. Land* **2013**, *5*, 511–520. [CrossRef]
6. Gao, H.; Wu, Z. Vegetation restoration and its effect on runoff and sediment processes in the Toudaoguai-Tongguan section of the Yellow River. *Acta Geogr. Sin.* **2021**, *76*, 1206–1217. (In Chinese)
7. Tian, F.; Liu, L.; Yang, J.; Wu, J. Vegetation greening in more than 94% of the Yellow River Basin (YRB) region in China during the 21st century caused jointly by warming and anthropogenic activities. *Ecol. Indic.* **2021**, *125*, 107479. [CrossRef]
8. Piao, S.L.; Wang, X.H.; Ciais, P.; Zhu, B.; Liu, J. Changes in satellite-derived vegetation growth trend in temperate and boreal Eurasia from 1982 to 2006. *Glob. Chang. Biol.* **2011**, *17*, 3228–3239. [CrossRef]
9. Jackson, R.B.; Jobbágy, E.G.; Avissar, R.; Roy, S.; Barrett, D.; Cook, C.; Farley, K.; Maitre, D.; Mccarl, B.; Cook, C. Trading water for carbon with biological carbon sequestration. *Science* **2005**, *310*, 1944–1947. [CrossRef]
10. Ji, G.; Song, H.; Wei, H.; Wu, L. Attribution Analysis of Climate and Anthropic Factors on Runoff and Vegetation Changes in the Source Area of the Yangtze River from 1982 to 2016. *Land* **2021**, *10*, 612. [CrossRef]
11. Calder, I.R.; Smyle, J.; Aylward, B. Debate over flood-proofing effects of planting forests. *Nature* **2007**, *450*, 945. [CrossRef]
12. Buttle, J.M.; Metcalfe, R.A. Boreal forest disturbance and streamflow response, northeastern Ontario. *Can. J. Fish Aquat. Sci.* **2000**, *57*, 5–18. [CrossRef]
13. Ceballos-Barbancho, A.; Morán-Tejeda, E.; Luengo-Ugidos, M.Á.; Llorente-Pinto, J.M. Water resources and environmental change in a Mediterranean environment: The south-west sector of the Duero river basin (Spain). *J. Hydrol.* **2008**, *351*, 126–138. [CrossRef]
14. Zhou, G.; Wei, X.; Luo, Y.; Zhang, M.; Li, Y.; Qiao, Y.; Liu, H.; Wang, C. Forest recovery and river discharge at the regional scale of Guangdong Province, China. *Water Resour. Res.* **2010**, *46*, W09503. [CrossRef]
15. Wang, S.; Fu, B.; He, C.; Sun, G.; Gao, G. A comparative analysis of forest cover and catchment water yield relationships in northern China. *For. Ecol. Manag.* **2011**, *262*, 1189–1198. [CrossRef]
16. Liu, Y.; Tang, X.; Sun, Z.; Zhang, J.; Wang, G.; Jin, J.; Wang, G. Spatiotemporal precipitation variability and potential drivers during 1961–2015 over the Yellow River Basin, China. *Weather* **2019**, *74*, 32–39. [CrossRef]
17. Yi, H.; Mu, X.; Peng, G.; Zhao, G.; Wang, F.; Sun, W.; Zhang, Y. Spatial Variability and Periodicity of Precipitation in the Middle Reaches of the Yellow River, China. *Adv. Meteorol.* **2016**, *2016*, 9451614.
18. Ji, G.; Lai, Z.; Xia, H.; Liu, H.; Wang, Z. Future Runoff Variation and Flood Disaster Prediction of the Yellow River Basin Based on CA-Markov and SWAT. *Land* **2021**, *10*, 421. [CrossRef]
19. Liu, W.; Shi, C.; Zhou, Y. Trends and attribution of runoff changes in the upper and middle reaches of the Yellow River in China. *J. Hydro-Environ. Res.* **2021**, *37*, 57–66. [CrossRef]
20. Hu, C.; Ran, G.; Li, G.; Yu, Y.; Wu, Q.; Yan, D.; Jian, S. The effects of rainfall characteristics and land use and cover change on runoff in the Yellow River basin, China. *J. Hydrol. Hydromech.* **2021**, *69*, 29–40. [CrossRef]
21. Liu, Q.; Yang, Z.F.; Cui, B.S.; Sun, T. Temporal trends of hydro-climatic variables and runoff response to climatic variability and vegetation changes in the Yiluo River basin, China. *Hydrol. Process.* **2009**, *23*, 3030–3039. [CrossRef]
22. Xu, J.; Jiang, X.; Sun, H.; Xu, H.; Zhong, X.; Liu, B.; Li, L. Driving forces of nature and human activities on water and sediment changes in the middle reaches of the Yellow River in the past 100 years. *J. Soils Sediments* **2021**, *21*, 2450–2464. [CrossRef]
23. Liu, X.; Liu, C.; Yang, S.; Jin, S.; Gao, Y.; Gao, Y. Influences of shrubs-herbs-arbor vegetation coverage on the runoff based on the remote sensing data in Loess Plateau. *Acta Geogr. Sin.* **2014**, *69*, 1595–1603. (In Chinese)
24. Liang, W.; Bai, D.; Wang, F.; Fu, B.; Yan, J.; Wang, S.; Yang, Y.; Long, D.; Feng, M. Quantifying the impacts of climate change and ecological restoration on streamflow changes based on a Budyko hydrological model in China's Loess Plateau. *Water Resour. Res.* **2015**, *51*, 6500–6519. [CrossRef]
25. Li, S.; Liang, W.; Fu, B.J.; Lv, Y.; Fu, S.; Wang, S.; Su, H. Vegetation changes in recent large scale ecological restoration projects and subsequent impact on water resources in China's Loess Plateau. *Sci. Total Environ.* **2016**, *569*, 1032–1039. [CrossRef]
26. Zhang, J.; Ma, X.; Li, Y. Vegetation dynamics and its impact on runoff in the Hekouzhen-Longmen region of the middle reaches in the Yellow River from 1980–2016. *South-North Water Transf. Water Sci. Technol.* **2020**, *18*, 91–109. (In Chinese)
27. Wang, X.; Yang, D.; Fang, X.; Cheng, C.; Zhou, C.; Zhang, X.; Ao, Y. Impacts of ecological restoration on water resources in middle reaches of Yellow River. *Bull. Soil Water Conserv.* **2020**, *40*, 205–212. (In Chinese)
28. Su, Y.; Chen, X.; Li, Y.; Liao, J.; Ye, Y.; Zhang, H.; Huang, N.; Kuang, Y. China's 19-year city-level carbon emissions of energy consumptions, driving forces and regionalized mitigation guidelines. *Renew. Sustain. Energy Rev.* **2014**, *35*, 231–243. [CrossRef]

29. Xia, H.; Wang, H.; Ji, G. Regional Inequality and Influencing Factors of Primary PM Emissions in the Yangtze River Delta, China. *Sustainability* **2019**, *11*, 2269. [CrossRef]
30. Fu, B.P. On the calculation of the evaporation from land surface. *Chin. J. Atmos. Sci.* **1981**, *5*, 23–31. (In Chinese)
31. Li, D.; Pan, M.; Cong, Z.; Wood, E. Vegetation control on water and energy balance within the Budyko framework. *Water Resour. Res.* **2013**, *49*, 969–976. [CrossRef]
32. Li, J.; Peng, S.; Li, Z. Detecting and attributing vegetation changes on China's Loess Plateau. *Agric. For. Meteorol.* **2017**, *247*, 260–270. [CrossRef]
33. Gao, J.; Jiao, K.; Wu, S. Investigating the spatially heterogeneous relationships between climate factors and NDVI in China during 1982 to 2013. *J. Geogr. Sci.* **2019**, *29*, 1597–1609. [CrossRef]
34. Guo, L.; Zuo, L.; Gao, J.; Jiang, Y.; Zhang, Y.; Ma, S.; Zou, Y.; Wu, S. Revealing the Fingerprint of Climate Change in Interannual NDVI Variability among Biomes in Inner Mongolia, China. *Remote Sens.* **2020**, *12*, 1332. [CrossRef]
35. Gong, L.; Jin, C. Fuzzy Comprehensive Evaluation for Carrying Capacity of Regional Water Resources. *Water Resour. Manag.* **2009**, *23*, 2505–2513. [CrossRef]
36. He, B.; Miao, C.; Shi, W. Trend, abrupt change, and periodicity of streamflow in the mainstream of Yellow River. *Environ. Monit. Assess.* **2013**, *185*, 6187–6199. [CrossRef]
37. Zeng, Z.; Liu, J.; Savenije, H.H. A simple approach to assess water scarcity integrating water quantity and quality. *Ecol. Indic.* **2013**, *34*, 441–449. [CrossRef]
38. Yan, D.; Lai, Z.; Ji, G. Using Budyko-Type Equations for Separating the Impacts of Climate and Vegetation Change on Runoff in the Source Area of the Yellow River. *Water* **2020**, *12*, 3418. [CrossRef]
39. Leuning, R.; Zhang, Y.Q.; Rajaud, A.; Cleugh, H.; Tu, K. A simple surface conductance model to estimate regional evaporation using MODIS leaf area index and the Penman-Monteith equation. *Water Resour. Res.* **2008**, *44*, 1–17. [CrossRef]
40. Liu, C.M.; Wang, Z.G.; Yang, S.T.; Sang, Y.; Liu, X.; Li, J. Hydro-informatic modeling system: Aiming at water cycle in land surface material and energy exchange processes. *Acta Geogr. Sin.* **2014**, *69*, 579–587. (In Chinese)
41. Zhang, D.; Liu, X.M.; Bai, P. Different influences of vegetation greening on regional water-energy balance under different climatic conditions. *Forests* **2018**, *9*, 412. [CrossRef]

 *land*

Concept Paper

Water Diversion in the Valley of Mexico Basin: An Environmental Transformation That Caused the Desiccation of Lake Texcoco

Carolina Montero-Rosado, Enrique Ojeda-Trejo, Vicente Espinosa-Hernández *, Demetrio Fernández-Reynoso, Miguel Caballero Deloya and Gerardo Sergio Benedicto Valdés

Colegio de Postgraduados, Campus Montecillo, Estado de México, Texcoco 56230, Mexico;
montero.carolina@colpos.mx (C.M.-R.); enriqueot@colpos.mx (E.O.-T.); demetrio@colpos.mx (D.F.-R.);
mcaballero@colpos.mx (M.C.D.); bsergio@colpos.mx (G.S.B.V.)
* Correspondence: vespinos@colpos.mx

Abstract: Mexico's basin is one of the most altered in the country, owing to the presence of the megalopolis of Mexico City. Lake Texcoco, which had the basin's biggest extension, dried up almost completely. The basin's evolution over time led in the formation of a megabasin in which water is transported from one source to another to serve the urban region and subsequently drained to prevent flooding. The major hydrotechnical works in Mexico Basin have been interpreted as a solution to the problem of flooding in Mexico City, but they were actually part of a much larger strategy of territorial appropriation by the Spanish colonists. The ecological imbalance that has resulted has sparked a variety of social issues. For the purpose of analyzing the environmental transformation of Lake Texcoco over the last 500 years, actors and processes that influenced specific moments in the country's history were identified; these elements showed the inexorable relationship between the lake and Mexico City. Subsequently, they were grouped by periods with similar trends in terms of the way in which society relates to and appropriates the natural environment of the lake. It was found that the critical moment for the desiccation of Lake Texcoco occurred during the Spanish colonial historical period as part of the redesign of the city; from then on, the same environmental imaginary prevailed century after century, shaped by social and economic factors. This study contributes to the literature on how urbanization affects natural resources by making an original theoretical contribution through an analysis based on political ecology, and it adds to the literature on how people use the prevailing federal area of the lake.

Keywords: Mexico City; historical political ecology; urban ecology; water spatial policy; environmental change

Citation: Montero-Rosado, C.; Ojeda-Trejo, E.; Espinosa-Hernández, V.; Fernández-Reynoso, D.; Caballero Deloya, M.; Benedicto Valdés, G.S. Water Diversion in the Valley of Mexico Basin: An Environmental Transformation That Caused the Desiccation of Lake Texcoco. *Land* **2022**, *11*, 542. https://doi.org/10.3390/land11040542

Academic Editors: Wiktor Halecki, Dawid Bedla, Marek Ryczek and Artur Radecki-Pawlik

Received: 15 March 2022
Accepted: 5 April 2022
Published: 8 April 2022

Publisher's Note: MDPI stays neutral with regard to jurisdictional claims in published maps and institutional affiliations.

Copyright: © 2022 by the authors. Licensee MDPI, Basel, Switzerland. This article is an open access article distributed under the terms and conditions of the Creative Commons Attribution (CC BY) license (https://creativecommons.org/licenses/by/4.0/).

1. Introduction

Lacking natural drainage, in the XV century, the Valley of Mexico Basin was an elevated basin of five interconnected lakes whose levels rose during the rainy season and fell during the dry months [1]. On an island in Lake Texcoco, the capital of the Aztec empire was founded. Through the construction of a complex engineering system, they handled flooding as well as drought while also assuring food production, urban water supply, and navigation [2]. Lake Texcoco originally spanned nearly the whole basin; it shrunk from a surface area of 7868 square kilometers to only 15.83 square kilometers [1].

After the Spanish conquest, the Spaniards took the indigenous Tenochtitlan as their own capital and strove to organize a regional economy modeled on Iberian patterns, which required more extensive modification of the basin's hydrology. Finally, in 1608, they decided to drain the Mexico Basin to the Tula River [3]. After the independence of Mexico from Spain, the draining continued until 1910, when the Grand Drainage Canal was inaugurated and the desiccation of Lake Texcoco was almost complete [4].

To understand the case of the desiccation of Lake Texcoco, it is necessary to analyze the sociopolitical actors and processes that generated or exacerbated the environmental problem in its long-term historical dimension. The purpose is to inform decision makers and improve environmental conditions while, at the same time, promoting nature conservation [5,6]. Bailey and Bryant [7] also mention that environmental problems cannot be fully understood in isolation from the political and economic contexts within which they originate.

The draining of Lake Texcoco is immersed in power relations and decisions about water resource management, which are reflected in historical conflicts over the use and management of lake resources in the Valley of Mexico. From the early Spanish colonial bureaucrats who sought to lower the surface area of the lake, to the nineteenth and twentieth century hydrological experts, the power groups that can be identified as elites in this region have long been at odds with nature. The objectives were frequently contentious which included the eradication of indigenous culture and customs along the lakeshore; the expansion of arable land for agriculture; the promotion of capitalist economic development; the improvement of sanitation; and the parceling out of lots to poor and working-class residents.

Political ecology and historical analysis of decisions that led to environmental changes are essential tools for understanding processes that have occurred over long periods of time [8–13]. The historical approach is especially important for understanding problems involving highly modified environments such as Lake Texcoco, where the natural environment is not just the backdrop for human events, it evolves on its own, both independently and in response to human activity [14].

Finally, the regional context of the society–environment relationship allows us to better understand the scale of a socio-territorial problem, the networks of actors involved and their discourses around it; it also allows us to identify the economic and political dynamics that have an impact on local resource degradation [12]. Figure 1 shows the four stages identified in this research by the policies applied. It can be seen that the critical moment in which the degradation begins occurs during the Spanish colony.

1325-1521 TENOCHTITLAN	1521-1810 SPANISH COLONY	1810-1910 INDEPENDENT MEXICO	1910-2020 MODERN MEXICO
Adaptation to the environment. Construction of agricultural and domestic artificial islands, canals, and dikes Flood control.	Environment adaptation Deforestation, urban redesign, cattle raising. Opening to prevent flooding	Political instability. Colonial environmental policies continue. Building a 50-kilometer long canal to drain the Mexico basin.	Drainage from Mexico City and Lake Texcoco is diverted to Tula basin. Environmental repercussions include differential sinking, desiccation of Lake Texcoco, aridification, and dust storms.

Figure 1. Identified stages for environmental policies in Lake Texcoco history. Author elaboration based on critical moments.

Lake Texcoco's history provides a window into the dynamism of the political economy, interactions between society and environment, and the numerous uses and competing developments. This research focuses on identifying the players and processes that contributed to the lake's desiccation throughout its history, with the historical processes grouped into four distinct stages defined by the political backdrop and management decisions made about the lake's environment.

2. Study Area: Former Lake Texcoco

The study area comprises the area of the former Lake Texcoco basin, which is part of the Mexico Basin located at the highest point of the Mexican plateau (2198 to 5200 m above sea level) with an average area of 9220 square kilometers. It was originally an endorheic basin surrounded by mountain ranges and volcanoes with interconnected lakes that functioned as a system of communicating vessels [1].

At its peak, Lake Texcoco had a surface area of 7868 square kilometers, occupying almost the entire basin. Over time, the lake divided into other lakes, creating a system of fresh and brackish water. In the pre-Hispanic period, the lake complex consisted of four large lakes: Zumpango, Xaltocan and Texcoco to the north with salt water, and Xochimilco to the south with fresh water. Lake San Cristobal, Lake Mexico and Lake Chalco were created by the construction of dams in the lakes of Xaltocan, Texcoco, and Xochimilco, respectively (ibid).

The origin of the soils in the ex-lake basin is related to the mechanical deposition of sediments from alluvial and lacustrine processes; their evolution has been conditioned to the parental material (sediment), the accumulation of organic matter, fine materials and carbonate, which resulted in the formation of entisols to mollisols or vertisols. Artificial drainage and desalination accelerated the pedogenic processes and allowed their incorporation into agriculture [15].

The study area's primary issues are significant environmental degradation caused by deforestation, lake surface reduction, soil erosion, loss of terrestrial and aquatic habitats, river contamination, and uncontrolled urban sprawl. Due to overexploitation and depletion of aquifers, which mostly affects the eastern portion of the Valley of Mexico, differential sinking has resulted in the extraction and pumping of water from the Lerma and Cutzamala rivers to meet Valley of Mexico demand, affecting external basins [16].

Historically, human settlements in the basin had an intimate relationship with water, and the lake's desiccation was closely related to the development of urbanization and Mexico City's expansion.

Overview of Existing Studies

Water historiography in the Valley of Mexico has been addressed by various authors, prioritizing immediate environmental problems, engineering or political aspects [4,17–22]. Candiani's work on the environmental evolution of Mexico City throughout colonial times [11] is an extremely intriguing and illustrative example. Her research focuses on one of the most difficult and pioneering engineering projects in the Mexico Basin, the Huehuetoca tunnel, which was designed to desiccate the lakes surrounding Mexico City. It investigates the social motivations underlying the drainage project's numerous structures and technological choices but is specific to colonial Mexico.

Iracheta and Dávalos [22] take a descriptive and historical approach to water in Mexico's basin, compiling a wide bibliography on themes such as the lacustrine system, flooding and drainage, hydraulic techniques, environmental sanitation, productive applications, and urban supply. Another historical descriptive approach is that of Lattes [19], who compiles the stories of personalities such as Spanish friars from the colonial era and discuss the floods in Mexico City and the hydrotechnical efforts conducted until the stage of Porfirio Díaz. He also includes historical maps that are critical for the study of drainage in the Mexico basin.

Boyer and Wakild [20] examine President Lazaro Cárdenas's six-year presidency (1934–1940) through the lens of environmental history. The Cárdenas administration is most recognized for implementing the revolution's "promises" through land reform, popular organization, and nationalization. Regardless, such statements obscure nature. Social landscaping, as they call it, was a key component of Cárdenas' ambitious social and political agenda. This is an intriguing approach, but it is again centered on a limited moment in Mexico's history.

Tortolero [21] wrote a historical analysis of water and environmental history in the Mexican basin in which he argues that the history of water management and use can serve as a lens through which to analyze diverse spheres of human activity. It is analyzed from four perspectives: the history of production and consumption, cultural history, economic history, and environmental history. This investigation is critical for gaining a better understanding of what has occurred in the Mexico basin.

In his work entitled *Political Essay on New Spain* [18], Alexander Von Humboldt describes the situation of natural resources in Mexico at the time of the colony. He points out that cultivation of morality can only be done in the groove of freedom; freedom, as a rule of balance in society, also allows the interactions of economic liberalism, noble actions of the new humanitarianism, anti-slavery attitude, and the free desire for the idea of progress. Humboldt frequently demonstrates the absence of freedom in New Spain and the resulting ethical–economic evils: authoritarianism, backwardness, morals, and a lack of culture. This is a seminal work for studies on the transformation of the landscape.

During the Porfiriato period, it was feasible to compile a report detailing all drainage work completed to that point [4]. It is a two-volume study that includes maps and extensive information about the state of the lakes prior to, during, and after the works are completed. While it compiles information about the social and political contexts in which the works were conducted, its primary contribution is technical and administrative data.

From the 1880s to the 1940s, Vitz [23] traces the failure of different conceptions for the engineering and restructuring of the Basin of Mexico, as well as the workers, peasants, and popular sectors that negotiated, fought, or attempted to benefit from these changes. An elite group of engineers, urban planners, and technocrats devised a strategy to push capitalist urban development on a rebellious populace and environment. According to this author, the hydraulic operations necessary to complete the drainage of the basin's lakes, new forest laws that deprived people of their community woods, changes in land use following the Mexican revolution, and the emergence of working-class colonies following 1920 are all part of the same story: Mexico City's capitalist rise.

However, Mexico City's floods and the drainage of the basin encompass a broader range of government policies, lake system descriptions, hydraulic techniques, productive uses, environmental sanitation, legislation, and litigation that can be better understood over a longer period of time than the ones mentioned above.

During the pre-Hispanic period, the dominant social discourse was about the scarcity of drinking water and the need for water conduction works for irrigation and food production; and at the same time, the need to prevent floods in Tenochtitlan [2,24–26].

For the Spaniards and the later independent government of Mexico (16th to 20th centuries), the lake was a problem due to the constant flooding in the city, the insalubrity and the *tolvaneras* (dust storms) caused by the gradual drying up of the lake. As a result, the decision was made to construct works that would allow free circulation of water and drainage of the city out of the Mexican basin towards the Tula River, which was completed in 1910 with the inauguration of the Great Drainage Canal, a work that accelerated the drying up of Lake Texcoco in the following decades [4,17].

3. Research Methodology and Theoretical Approach

In this study, a historical political ecology approach is used, in which an adaptation of Blaike and Brookfield's decision-making in land management degradation [9] is proposed as a tool to identify the political–environmental relationship, exposing different environmental policies regarding Lake Texcoco management and grouping them by periods (see Scheme 1).

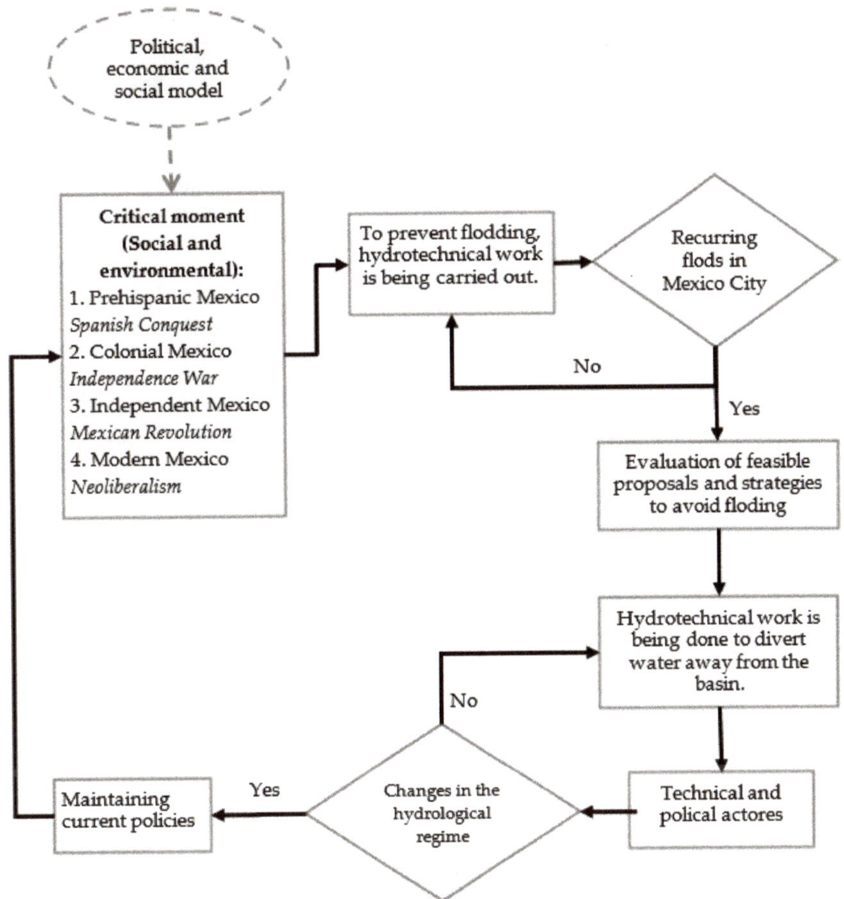

Scheme 1. Decision-making in historical Lake Texcoco for alterations to the hydrological regime.

Political ecology originated in the 1980s as an interdisciplinary topic based on the core concept that ecological change cannot be understood without taking into account the political and economic factors that surround it [27]. There are several approaches within this discipline depending on how the notion of power is applied to understand human/society/nature connections. In this research, the environment is regarded as "power-laden rather than politically inert" [28].

According to Zimmerer [29] methodology design and selection of methods should be carried out simultaneously, particularly with the selection of the theoretical framework and conceptual concerns that will be explored. In the case of social–ecological change, such methods must be compatible with the processes, spatial scales, and temporal frames that constitute the study objective.

For social–ecological change, Zimmerer (ibid) proposes spatial scales and temporal framing within discourse and content analysis. To delimit specific timeline stages in this study, two concepts related to environmental change were used as starting points for analysis: critical moment and episodic events.

The first concept has been applied in participatory nature conservation projects [30] and is described by Khan [6] as a:

"Conspicuous and sensitive moment that offers a specific insight into the interplay and autonomy of the actors involved in an event, one that illustrates or informs a political ecology analysis. The critical moment exposes the interplay and autonomy of both endogenous and exogenous actors" (p. 1)

The second concept is mentioned by Bailey and Bryant [7] in historical political ecology analysis to understand politicized environments in terms of physical changes, rate of impact, and political response; the episodic dimension is one that is often sudden and can be prolonged, impacting society in general but with unequal damage exposure, and the political response is focused on disaster relief.

Questions that led this study were:

- Who held power at different times in history?
- What were the mechanisms and instruments used to legitimize power?
- What were the objectives pursued?
- What were the decisions taken?
- What were the environmental consequences?

A review of little-known documentary evidence and historical geographic information on the surface of the lakes, drainage infrastructure, and urban growth was conducted. The data used in this study include archival material from technical documentation about drainage in the Mexico Basin, as well as environmental and socio-political descriptions from each identified stage; the reviewed materials are mainly Spanish-language materials and were first published in Mexico.

In each historical period analyzed, different hydrotechnical works were carried out to prevent the flooding of Mexico City. The type of works built were based on the political moment, the purposes and interests of the actors, the economic resources or the state of emergency in the face of an environmental catastrophe. The impact of this was the reduction in the lake area until it dried up, which in turn allowed the expansion of Mexico City on the uncovered lands of the lake, while the deepest part of the lake became a federal area with exposed saline soils and small artificial lagoons that capture runoff from the basins to the east of the basin.

Limitations of the Study

Given the extension of the time period studied, information on some other factors that influenced the desiccation of Lake Texcoco may have been inadvertently omitted. Mexico has gone through several moments of transition and change, some of them violent social movements that could reduce the veracity and availability of the information at present; also, there was limited availability of data on the oldest events that were studied.

4. Processes and Actors That Influence the Desiccation of Lake Texcoco

4.1. Tenochtitlan and Lake Texcoco (1325–1521)

According to historians, the modern state of Mexico began in the late postclassic period, 625 years ago, with the founding of Tenochtitlan in 1325. Tenochtitlan's founders, the Mexica, were the last Mesoamerican people to build the city in the fourteenth century on one of the western islets of Lake Texcoco in the Valley of Mexico basin [25].

For two centuries, the Mexica altered the lake system, resulting in the development of intricate hydraulic infrastructure systems. The works served a variety of purposes and were constructed on both land and in riparian and lake environments. Palerm [24] identified and defined the most significant hydrotechnical works including *Chinampas* (described by the author as an "inside lagoon", an artificial island built by the Mexica for agriculture and settlement), causeways, *albarradones* (colonial and pre-Hispanic hydraulic works that functioned as dikes), flood defense and drainage works, freshwater conduction through canals, ditches and aqueducts, and formation of lagoons and artificial swamps.

At the middle of the island was the ceremonial center. The population settled around it on chinampas where they built their houses and developed agriculture, all this without damaging its natural lacustrine condition. The causeways that linked the island to the mainland in turn divided the lake. They also built aqueducts to supply drinking water to the island from the springs of Chapultepec and Coyoacán, navigable canals for the transport of people or goods and ditches for the distribution of drinking water and irrigation.

Tenochtitlan did not escape flooding. The community that evolved within that lake ecosystem figured out how to manage the lakes' water levels in order to limit damage. According to historical sources, two main constructions had their origin in environmental events, the Nezahualcóyotl and the Ahuízotl *albarradón* [31].

As a result of the 1449 flood, Nezahualcóyotl (*tlatoani* or emperor of Texcoco from 1402 to 1472) built a 16-km-long *albarradón* that connected the Tepeyac hill to the Santa Catalina Mountain range in order to prevent flooding caused by Lake Texcoco's overflow. This also allowed for the segregation of the Mexico Lagoon's fresh waters from the saline waters of Lake Texcoco.

On the other hand, the flood of 1499 was triggered by Ahuízotl (Tlatoani of Tenochtitlan from 1486 to 1502) when he attempted to build an aqueduct connecting Coyoacán and the city. The city was flooded for 40 days following the inauguration of the aqueduct and, in response to this catastrophe, they built another *albarradón*.

The Mexica developed a great empire in Mesoamerica through hydrotechnical works; by the 16th century, the city of Tenochtitlan had a population of between 170 thousand and 200 thousand inhabitants, and the entire basin had a population of about 1 million [2]. The prosperity of the cities inside the basin demonstrates that they used water as a tool to achieve a successful empire.

4.2. The Spanish Colony and the Drain to Prevent Flooding (1521–1810)

During the period of the Spanish conquest, there was a significant change in the landscape of Lake Texcoco. It is at this point in history that there is a transformation in the way people relate to the lake environment.

The Spanish explorers and conquistadors led by Hernán Cortés entered Tenochtitlan on 8 November 1519. After almost two years of invasion, the final strategy of the Spaniards was to cut off the supply of drinking water and the entry of food. The Mexica were defeated by using the same natural resource that they had used to establish their empire.

What occurred following the conquest would be significant for the future of Lake Texcoco. The capital of New Spain had to be built, and among the possible locations were Coyoacán, Tacuba, Texcoco, and Tenochtitlan. Hernán Cortés is credited with the idea to establish Mexico City on the ruins of Tenochtitlan, citing political measures to ensure the Spanish conquest was effective and definitive. In this regard, José María Marroquí in his book *La Ciudad de México* published in 1900 [32], mentions that Cortés considered it dangerous to leave the old city free given the possibility that over time the indigenous people could try to recover their temples, palaces and monuments.

Between the demolition of the Mexica city and the creation of the Spanish one, forty years passed. New emblematic buildings were built over the old ones such as the Cathedral of Mexico over Templo Mayor ruins. The pre-Hispanic period moved gradually towards the consolidation of Spanish rule and the beginning of the colonial period. Historians date the end of the conquest stage to around 1560, when tribute began to be expressed in pesos and paid in currency [33].

The new layout of the city broke the natural balance with the lake environment that the Mexica maintained. Furthermore, forests were cut down to use wood for new buildings, to make furniture or as fuel. Furthermore, some rivers were diverted, dried up or polluted. All this altered the physiognomy and productivity of the basin, starting the process of change in the hydrological system [34].

As geographer and naturalist Alexander von Humboldt pointed out in his 1827 publication Political Essay on New Spain [18], "the first conquerors wanted the beautiful valley of Tenochtitlan to resemble in everything the Castilian soil, in the arid and devoid of its vegetation" (translation from Spanish). His observation tells us that, unlike the Mexica civilization, the Spaniards were unfamiliar with floodplains as inhabited spaces, and therefore the conquest altered the interaction between humanity and the basin's natural lake environment.

The changes they made to the pre-Hispanic hydraulic system caused major floods, prompting a series of decisions to mitigate them, including attempts to replicate some of the works built by indigenous people in the past, but without achieving the same results due to the change already present in the environment.

An example of this is the *albarradón* of Nezahualcóyotl, which was demolished to allow the entry of Spanish vessels, causing a great flood in 1555. The Viceroy of New Spain, Luis de Velasco y Ruz de Alarcón, ordered the construction of a dike between the Calzada de Guadalupe to the north and the Calzada de Iztapalapa to the south of the city to substitute for the Netzahualcóyotl *albarradón*. This construction would be known as the *albarradón* de San Lázaro [19].

However, the *albarradón* of San Lázaro was not enough to stop the floods. In 1580, another great flood occurred. At that time, Viceroy Martin Enrique de Almanza was overseeing New Spain, and the idea of draining the lagoon surrounding Mexico City started to be mentioned as a solution to flooding.

Another flood occurred in 1604 and another shortly after, in 1607. Viceroy Luis de Velasco y Castilla ordered the construction of a tunnel that would pass through the northeast corner of the basin with the purpose of draining Lake Zumpango and preventing further flooding. The design would divert the waters of the Cuautitlán River to the Tula River. The person in charge of the design and construction of the Huehuetoca tunnel was the engineer and cosmographer Enrico Martínez [35]. The Valley of Mexico basin ceased to be an endorheic basin as a result of the development of this project.

Perló Cohen and González Reynoso [36] point out that the Huehuetoca tunnel prevented Mexico City from flooding as a consequence of the north's growing rivers and lakes. However, the east, south, and central valley overflows into Lake Texcoco were not controlled, and flooding occurred again in 1615 and 1623.

The flood of 1623 occurred during the government of Viceroy Diego Carrillo de Mendoza y Pimentel, who, according to Lattes [19], had opted to allow the lagoons to once again receive the flow of the rivers due to the high cost of maintaining the works. The entrance of the artificial drainage was blocked so that the waters of Zumpango could reach Lake Texcoco; the result was a gradual increase in water levels in the city and, by 1627, many of the streets were again flooded.

Numerous repairs were required both within the city and on the dikes and canals. The Huehuetoca tunnel was modified to become an open pit, allowing the necessary effluent to flow, but the Nochistongo gorge would not be completed until 1789 [35].

However, the most devastating flooding of the time had not yet occurred. In 1629, a deluge known as the San Mateo Deluge struck Mexico City. It rained continuously for 40 h. For five years, the city was submerged in water due to the flood. Water levels in the city center were recorded as high as two meters, and it would take until 1634 for a drought to put an end to such an overwhelming flood. The cost of the destruction was enormous. At the time, the population was believed to be 150,000, of which about 30,000 died as a result of the flood and its effects, such as the city's insalubrity. Another significant portion of the population emigrated to various cities throughout the country.

Throughout the 18th century, the government of New Spain continued to work on the drainage of the Valley of Mexico basin. Among them were the *albarradón* de San Mateo canal in 1747, the Guadalupe canal in 1796, and the Zumpango canal in 1798 to connect them to the Nochistongo gorge; despite the large lake area drained as a result of these works, the city continued to flood.

Valek Valdés [35] recounts the observations of the renowned geographer and naturalist Alexander von Humboldt on the works that supported the security of the capital at the beginning of the 19th century; among them were the stone causeways that prevented the waters of Zumpango from reaching the Lake San Cristóbal and ending in Lake Texcoco, the causeways and locks of Tláhuac and Mexicaltzingo, in the Nochistongo gorge that took the waters from Cuautitlán River to Tula River and, finally, the two Mier canals that allowed lakes Zumpango and San Cristóbal to drain.

As of now, the administration of natural resources was influenced by the omission of efforts that the country's native inhabitants had made for generations, shifting the concept of a civilization to adapt to the environment to now adapting the environment to the goals of the new civilization. According to Mann [26] in his analysis of the arrival of Spanish in the Americas, the assumption was that the new continent was pristine, and natural adaption mechanisms ignored.

4.3. The Drainage Proceeds during the Independent Period (1810–1910)

The country's political instability at the turn of the nineteenth century was evident in the condition of decay of the drainage works. During this time period, the government transitioned from a monarchy to a centralist republic to a federalist republic; hence, the drain was abandoned amid armed revolutions and power struggles. According to the engineer Francisco de Garay [3], it was forgotten at those times that personnel guarded the Nochistongo gorge in the north.

Dr. José María Mora, one of Mexico's most illustrious liberalism figures, was commissioned in 1823 to conduct a diagnosis on the state of the drainage work, in which he describes the state of complete abandonment and the pressing need for repairs. Mexico City was in danger of flooding due to the unevenness in relation to the lake caused by the constant deposit of earth carried by the rivers that flowed into the lake. Additionally, he expresses his opinion on the mishandling of past works, which were pricey and brief in duration [37].

In 1847, as the North American army marched towards the Valley of Mexico and Antonio López de Santa Anta was president of the Republic, the eastern sector was flooded as a defensive measure, digging the Viga canal and breaching the Mexicaltzingo locks. This move failed as a military tactic and instead served to exacerbate the risk of flooding in the capital and southern villages. Following that, engineers Francisco de Garay and M.L. Smith repaired the structure.

On 22 April 1853, the Ministry of Development was established, having authority over the canals and drainage of the Valley of Mexico. The proposal to use the city's hydraulic system for navigation was made as early as 1830 by Lucas Alamán, the country's Minister of Interior and Foreign Affairs Relations at the time, and was implemented briefly, but the shallowness of the canals and dikes, as well as the maintenance costs associated with them, increased interest in drainage. [4].

It was not until 1856 that the waters of Lake Texcoco threatened the city again, and in accordance with the established plan of action, a thirty-member General Board was appointed, which would later appoint a Minor Board to oversee the necessary repairs to the city's aging hydrotechnical works, with engineer Manuel Gargollo assigned to the north section, Juan Manuel de Bustillo to the center, and Francisco Garay to the south.

To the north, the Tepotzotlán and San Ignacio outlets were restored to the west bank of the Cuautitlán River, canals were cleaned to restore water to Lake Zumpango, and the Guadalupe canal was reopened. The San Cristóbal dike was reinforced in the center, and it was proposed that the Teotihuacan dike be restored in order to unload the Texcoco reservoir. It was, nevertheless, completely planted and occupied. The landowners presented the solution to facilitate the deposit of the waters by erecting many dams. All of these efforts were insufficient due to their unplanned nature, but in the south, the engineer Garay's efforts were part of a scheme presented years previously that was more successful.

In the south, the diversion in the Mexicaltzingo dike was closed, and the breaches in the Culhuacán and Tláhuac causeways were fixed, while the walls' height was increased to compensate for being totally submerged by the lakes' water. Additionally, the San Lorenzo canal was constructed to cut through the dividing hill between Xochimilco and Texcoco, and a floodgate was constructed in Mexicaltzingo to simply and rapidly manage water flow. Over the next five months, the Chalco and Xochimilco lakes increased their water levels by 56 cm, while the Texcoco lake decreased.

"The lake of Texcoco being the lowest in the entire valley and with no outlet for its waters, all the salts that the landslides of the mountains pull on their route are deposited there," said Francisco de Garay [3]. Additionally, the salting in lakes was becoming more evident because in the dry season, during the ebb of its waters, salts sprout from the ground, rising to the surface by the capillarity effect or evaporation. The sun evaporates the water, leaving the salt behind, causing the deposit to grow (ibid).

De Garay also mentions that in the time prior to the Spanish conquest, it was common to find a large quantity of fish in Lake Texcoco, but that it had disappeared for at least a century. While the *albarradón* de Nezahualcóyotl was standing, the rivers of the west flowed around the capital, and through the San Lázaro floodgate, there was a freshwater stream at the foot of Peñón Chico, which turned it into a beautiful orchard. With less water coming from the south and no water coming from the north, an important area lost its ability to grow food. Instead, tequezquite, a mineral salt, or ground salt, was harvested.

During his exploratory trips to the American continent in 1803–1804, Alexander von Humboldt [18] describes the distance from the center of Mexico City to Lake Texcoco as 4500 m, and more than 9000 m to Lake Chalco, and notes a depth of 3 to 5 m in some areas, and up to a meter in others.

Thus, during the first two-thirds of the nineteenth century, the valley's drainage and the drying out of Lake Texcoco were effectively halted, but once a degree of stability was achieved toward the end of the century, the drainage was once again promoted primarily by political actors and social elites, despite warnings about the potential environmental consequences that were already looming. As Miranda Pacheco [38] points out, even though drainage works were restarted in the 1870s, the city's long-term growth was due to the "interests and projects" that had been built on private property and public land since the beginning of the century, even before the country became independent (59 p.).

On the other hand, Vitz [23] notes that the environmental practices that dominated the twentieth century were patrimonial and commercial exploitation of natural resources, which benefited from political instability and a colonialist-influenced society. This occurred in a political climate in which decisions affecting the natural environment were made on the basis of what was urgent and convenient. He also emphasized the differences in how different socioeconomic classes relate to the environment.

4.3.1. The Decision That Perpetuated the Desiccation of Lake Texcoco in the Second Mexican Empire

For the first time during the Second Mexican Empire, drainage of the entire valley was considered to combat flooding in the city. Despite Maximilian of Habsburg's liberal thinking and appreciation for Mexican nature, the draining decision was influenced by the conservative group that participated in the monarchy's formation. Miranda Pacheco [38] highlights the participation of politicians, doctors, lawyers, engineers, geographers and the upper class of society.

Shortly after Maximilian of Habsburg's arrival in 1864, the city faced another flood, and to find a solution, national and international experts were invited. The board of experts chose engineer Francisco de Garay's proposal, which included a network of staggered canals for drainage, navigation, and irrigation, with north and south extensions. The objective was to be able to control the level of nearly all vessels by creating additional reservoirs and eradicating stagnant waters. It was a very ambitious project that required a significant expenditure, which is why it was harmed by budgetary constraints following the restoration of the Republic in 1867.

Maximilian chose as his first option the drainage plan proposed by engineer M.L. Smith in 1848, who recommended not drying the lakes but connecting them via a canal, where it would be possible to balance the lakes' waters by utilizing the difference in their heights, thereby avoiding overflow into the city. Others proposed the construction of artificial containers to repurpose excess water in agriculture. The order of 27 April 1866 reflected this decision; nevertheless, on 13 November the same year, the decree was published to carry out Francisco de Garay's plan, originally offered in 1856 and selected as the winner of the Minor Drainage Board's call for proposals.

This change of decision on the part of Maximilian was partly influenced by his advisers, who convinced him of the obstacles that the waters would present, overflowing or not, to his urban projects, as well as for those people close to him [38]. A letter written on 3 July 1866, by the First Engineer, Miguel Iglesias of the Ministry of Development, explains a different situation to Maximilian:

"Indeed, for our lovely capital, it is a benefit of such transcendence (the drain) that it may be stated that its life is in it, because without it, we will always have it nearly flooded and in risk of becoming completely flooded, and thus seeing it in ruins. It is also an issue for the other populations that dwell in its shadow: how much would their inhabitants lose if that center of business and consumption provided all of their needs? Additionally, even if it were not to protect them from the calamity of a flood, the welfare and health of the valley's inhabitants; the increase in fertility and production that will result from the cultivation of the vast lands currently occupied by the waters of one of the lakes that are so harmful due to the poor conditions in which they are found; the new and convenient communication routes created by the numerous canals that will cross the valley; all of these are enormous benefits" [4] (translation from Spanish)

Notably, M.L. Smith had previously emphasized the dry fields of the San Lázaro canal, an ancient link between Texcoco and Mexico City. He designated them as sterile plains due to their location within a salty lake basin. The flood of 1865 compelled them to act immediately, and as a result, a series of steps for the drainage works were undertaken. Francisco de Garay was appointed "sole and responsible director and inspector of all activities relating to the water issue in the Valley of Mexico" during this conflict. Francisco de Garay, on the other hand, was a devout Republican liberal who accepted the designation without title or compensation.

The lake's water had already reached the streets of Mexico City by October 1865. The situation was critical, and the city's engineers were called in to help. Francisco de Garay [3] indicated that the flood exceeded twice the usual surface of the lakes, invading the entire eastern part of the city, and that "to save the capital from danger, it was necessary to lower the level of Lake Texcoco". The urgent drainage plan presented by Francisco de Garay obtained most of the votes and was defended against Maximilian when he questioned its effectiveness. De Garay was so sure that he bet his own head [3].

The proposal called for isolating Lake Texcoco so that the northern and southern lakes would not continue to discharge their surpluses into the basin's lower reaches. Dikes were built, and help was once again sought from ranches, farms, and adjoining areas in order to construct artificial basins and reduce the risk of flooding in the city's towns and neighborhoods. The water level in Mexico City's historic center was 54 cm lower than in 1629. When the waters receded in 1866, the floors of the flooded streets and squares were ordered to be raised.

In May 1866, Emperor Maximiliano established the Drainage Commission, which included engineers Miguel Iglesias, Aurelio Almazán, Manuel Álvarez, and Jess Manzano, to conduct field studies and verify that the drainage project proposed by Smith was feasible, which they endorsed; work began at the end of the year. He also arranged for Miguel Iglesias to travel to Europe to purchase the necessary machinery [4].

The works conducted during this time period are detailed in the 1868 Development Memory and in the compilation of reports from the Historical, Technical and Administrative Memory of the Valley of Mexico Drainage 1449–1900; nevertheless, the lack of resources meant that most of the works of the drainage were postponed, and government engineers were then mainly concerned with conducting field studies and conservation works.

Maximilian of Habsburg was a European liberal who, upon his arrival, developed a policy in disagreement with the traditional position of the conservative class and the

Mexican clergy but, due to his short duration in government, his influence did not reach the entire national territory.

Mexico City's architectural design, which began at this time, would last for decades, and projects for lake drainage would also continue. Vitz [23] points out that during this time, the priority was to safeguard the city from floods and protect private property, even when the interdependence of these with their environment was recognized.

4.3.2. The Porfiriato (1876–1911) and the Basin's Second Opening

The hydraulic work that offers a second outlet to the basin is known as the Grand Drainage Canal. This work was part of several urbanization initiatives in Mexico City under the presidency of Porfirio Díaz, with the distinction that it was begun during Maximilian's reign. During the Porfiriato, Mexico experienced rapid development, which was made possible by foreign investment. The infrastructure of the railway, telegraph, telephone, and electricity were expanded. Despite the city's economic expansion and modernization, it was a time marked by economic disparities among the inhabitants.

The drainage works were cut off from the Ministry of Development in 1867 by the Ejército de Oriente in Mexico City. It was left behind because it was considered as an imperial endeavor. Andres Almazán, José Iglesias, and Jose Manzano came to Porfirio Díaz to ask for help. Porfirio Díaz's reaction, in his letter of 11 May 1867, is as follows [4]:

"Having read with pleasure VV's report on the work being done in Zumpango to aid Valley Drainage [..], I could wish for a few glories in my temporary position, such as adding momentum to these operations [...] As a result, [...] I have directed that the Federal District Treasury Department minister to them the sum of 1500 pesos each month for the conservation of drainage works, while the Supreme Government determines that they continue and carry out with appropriate diligence" (translation from Spanish)

Blas Balcárcel was heading of the Ministry of Development when the Republic was restored in 1867. Together with the Minister of Finance, José Mara Iglesias, he was successful in obtaining approval for a special tax to fund the drainage works. This tax included a 50% rise in municipal taxes collected at customs in the capital, as well as a 20% increase in direct taxes collected in the Valley of Mexico.

The drainage works were restarted in April 1868, but a later decree prohibited the special funds, and these entered a common fund. As a result, the works received an allocation from the general budget. The instability in the public peace influenced the perception of the amount destined for the works, which caused them to be paralyzed in October 1871. The completion of the outlet was the focus of the work done between 1868 and 1871.

The valley's drainage was also viewed as a solution to the city's health and hygiene issues. The Mexican Society of Geography and Statistics issued an opinion on the valley's drainage in 1874, arguing that doing so would prevent the physical and moral degeneration of their race and that the desiccated lands could be sold to benefit commerce, agriculture, hygiene, and the public treasury [38].

Miranda Pacheco [38] reveals the many perspectives on the valley's draining at the time. The Hygiene Commission of the Academy of Medicine in Mexico (at first), Doctor Eduardo Liceaga (president of the Medical Congress, family doctor, and close friend of Porfirio Daz), and Francisco de Garay were among those who supported it.

Those who opposed the drainage included Manuel Orozco y Berra, who believed that the city's drainage should be separated from the valley's drainage so that the Texcoco lake did not become a gigantic sewer. To stop the putrid emanations towards the city and limit the surface of the basin, Leopoldo Río de la Loza recommended reforestation of the area near Lake Texcoco. The doctor José G. Lobato noted, in his 1877 study, Hygienic Study on the Exanthematous Typhus, that preserving the surface of the lakes would reduce outbreaks of typhoid infections, a phenomenon that in pre-Hispanic times did not occur, but the Matlazahuatl pandemic happened when the lake's surface had dropped with the Spanish [38].

Rivalries between drainage officials, particularly between engineers Francisco de Garay and Luis Espinosa, were not new to the project. The tunnel's placement and hydraulic flow were two of the most contentious issues. Initially, the tunnel was designed to follow M.L. Smith's plan via the Acatlán ravine, and construction began in that direction. Francisco de Garay, on the other hand, thought it should go through the Ametlac ravine. When Manuel Fernández de Leal took over the Ministry of Development and designated him acting director of Works, Luis Espinosa's decision was eventually imposed.

As a result, until 1879, it was unclear how much water had to be extracted. The engineers predicted sizes and slopes for unusual and severe rain, which raised the cost of the project. Previously, M.L. Smith estimated a flow of 8 cubic meters per second, Francisco de Garay calculated a flow of 33 cubic meters per second, and Miguel Iglesias (head of the commission that studied and laid out the drain in 1866) calculated a flow of 41 cubic meters per second [4].

Porfirio Díaz accepted the drainage scheme for the Valley of Mexico presented by Luis Espinosa in 1879. It was based on the previous one by Francisco de Garay and took into account the work already completed by Miguel Iglesias but rectified the tunnel's hydraulic flow by 21 cubic meters per second. From 1871 to 1886, Luis Espinosa served as director of drainage works in the Valley of Mexico and as director of the Drainage Board of Directors.

According to Luis Espinosa [39], the lake of Texcoco was destined to disappear since the waters that came from the springs, wells, and drainage of the city were not enough to maintain the extension of the lake. Furthermore, he mentions the health issue commenting that "there are respectable opinions that foresee a danger to health in the complete drying up of the lakes".

With the return of Porfirio Díaz to power in 1884, the project gained enough momentum to be completed. The entire project consisted of three major parts: an open gorge 39.5 km long from Lake Texcoco to the northeast end of Lake Zumpango, a tunnel almost 10 km long from the shore of Lake Zumpango traveling northwest up to the Acatlán ravine, also known as Tequixquiac, and a canal. This would carry water from the Tequixquiac River, which meets the Tula River and, downstream, creates the Moctezuma River, to Pánuco, and finally empty into the Gulf of Mexico [4].

During the Porfirio Díaz administration, a budget close to 16 million pesos was allotted for the construction of the Grand Drainage Canal between 1886 and 1900. It was dedicated on March 17, 1900, in the presence of the President of the Republic, some Secretaries of State, various members of the diplomatic corps, the Drainage Canal Board of Directors, engineers, employees, and visitors from banking, commerce, industry, and the arts. In this regard, Luis Gonzales Obregón, a member of the Drainage Canal's Board of Directors, wrote:

"March 17, 1900, will be a memorable date, because the works inaugurated on this day together with those of sanitation, will make Mexico one of the most pleasant mausions, among the capitals of the American Republics, for its beauty, health and climate" (translation from Spanish)

Mexico City was already far removed from its lacustrine origins at the time of the Grand Drainage Canal's opening. The Hydrographic Commission of the Valley of Mexico estimated that the lakes' maximum extension during the rainy season would be 536.50 square kilometers, with the surface area of Lake Texcoco ranging between 183.28 and 238.54 square kilometers, depending on the amount of water received and the amount of precipitation expected throughout the year (Table 1).

Table 1. Lake surface on the Valley of Mexico in 1860—1870.

Lake	Area (km^2) Regular Season	Area (km^2) Rainy Season
Texcoco	183.28	272.17
Zumpango	17.20	21.70
Xaltocan	54.07	54.07
San Cristóbal	11.03	11.03
Xochimilco	47.05	63.36
Chalco	104.48	114.17
Total	417.11	536.5

Source: Luis Espinosa, Technical and administrative report on the Valle de México drain, 1902, in Legorreta, Jorge (2006). Estimates made with data from the Valle de México Hydrographic Chart, by Manuel Orozco y Berra.

In the act issued on the inauguration of the drainage works in the Valley of Mexico, the Grand Drainage Canal is described as a great and beneficial work that will liberate Mexico from floods and improve the hygienic conditions of the capital and the valley [4]. The work was seen as an engineering feat and a symbol of the progress and modernity of the country.

However, only four months later, a second flood revealed that the work was ineffective. The Grand Canal's status would deteriorate over time, as the canal's movement was determined by gravity and Mexico City was sinking. The Grand Drainage Canal displaced about one million cubic meters of water in its first four years [40].

Although the floods continued, this project was critical for future urban expansion because it created new territory that could be salvaged by draining the waters of Lake Texcoco. The decrease in Lake Texcoco to its current extent and the urban area of Mexico City are depicted in Figure 2.

4.4. Drainage of Mexico City in the Twentieth and Twenty-First Centuries: The Battle against the Lake Is Not Over

Efforts undertaken centuries ago to minimize flooding in Mexico were insufficient. The rapid growth of the city, constructed on a historic lake complex, prompted the development of increasingly complicated infrastructure. This would eventually impact the city's hydraulic infrastructure, necessitating ongoing drainage projects and repair of current ones. Reyes [41] writes about it:

"It covers the drying up of the valley from the year 1449 to the year 1900. Three races have worked in it, and almost three civilizations —that there is little in common between the viceregal organism and the prodigious political fiction that gave us thirty years of august peace. Three monarchical regimes, divided by parentheses of anarchy, are here an example of how the work of the State grows and corrects itself, in the face of the same threats from nature and the same earth to dig. The slogan of drying the earth appears to have run from Nezahualcóyotl to the second Luis de Velasco, and from him to Porfirio Díaz. Our century found us still throwing the last shovel and digging the last ditch"

(translation from Spanish)

The first Tequixquiac tunnel, inaugurated in 1900 by Porfirio Díaz, had a flow of 16 m 37 s; this was enough for Mexico City, which at that time had around 500,000 inhabitants and when the rivers that crossed it emptied into Lake Texcoco. Subsequently, the water discarded by the drainage system began to be used for irrigation of agricultural land, forming part of the irrigation district of Tula, Hidalgo in the Mezquital Valley region.

Only 37 years after the tunnel's completion, it became insufficient to accommodate Mexico City's population increase. By 1940, the city had grown to 1,757,530 inhabitants, and fearful of the first tunnel collapsing, the federal authorities decided to expand the system. Drainage was undertaken via a second tunnel through Ametlac ravine in Tequixquiac. The new tunnel would be the basin's third artificial outlet.

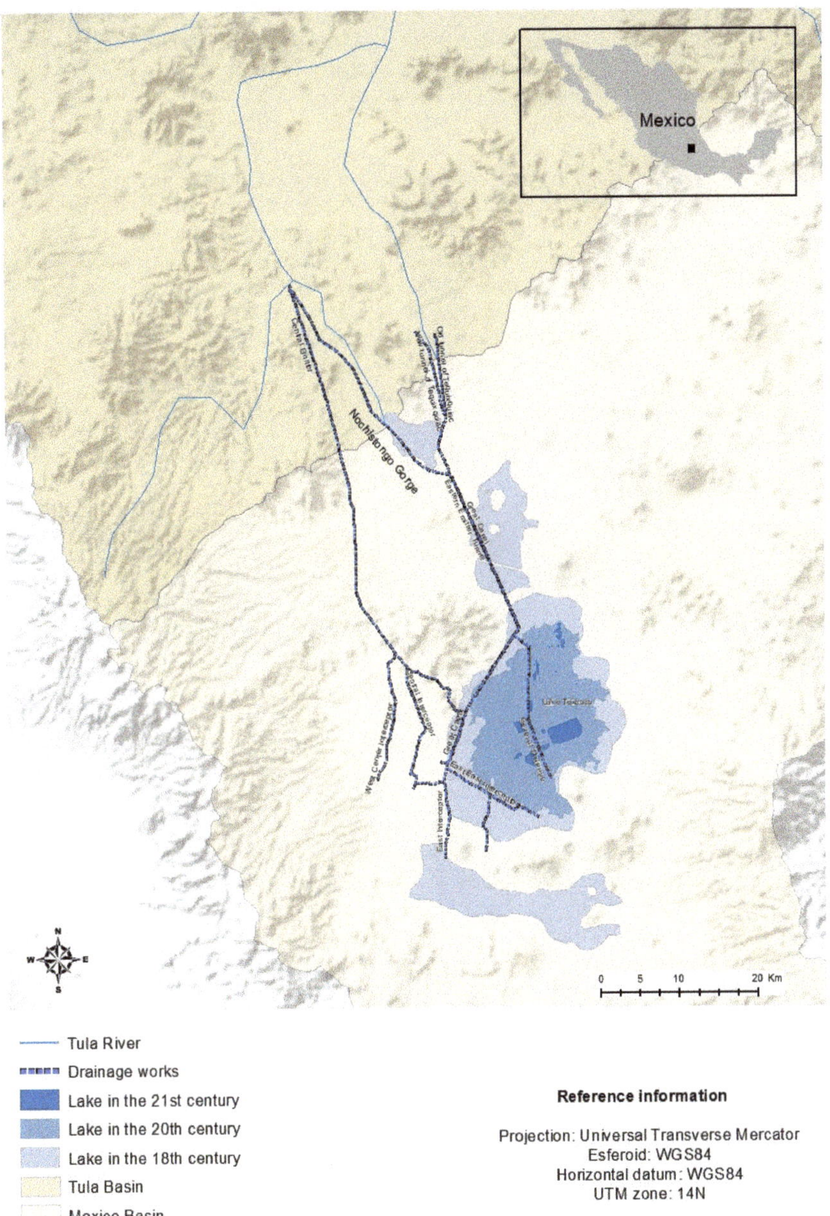

Figure 2. Hydrotechnical works that open Mexico Basin to Tula Basin. Own elaboration based on data from CONABIO [42] and Stangl [43], historical maps recovered from Mapoteca Orozco y Berra [44], and photo-interpretation of satellite images from 2021.

The second tunnel began construction on 1 July 1937, initially under the supervision of the Ministry of Communications and Public Works and later transferred to the Ministry of Hydraulic Resources. It was opened in 1946 by President Manuel Avila Camacho and had a flow rate of 60 cubic meters per second, a diameter of 4 m, a length of 11.3 km, and a slope of two thousandths.

However, there was a tremendous ecological cost. These three exits contributed significantly to drainage of the lakes and rivers, hence reducing the risk of floods in the city. The loss of the natural balance of the hydraulic system occurred as a result of the artificial outlets. Since 1608, the drainage flow has increased without being matched by an equal amount of incoming water, compressing the subsoil clays and resulting in differential subsidence in the soil.

The sinking has impacted not only Mexico City, but also other areas of the old lake of Texcoco, with one of the repercussions being the shutdown of the drainage system. When the Grand Drainage Canal was launched, the city's base was 5 m higher than it is now, the collectors had a sufficient slope, and the flow could be dragged to the Tequixquiac tunnel, but by 1950, the situation had changed. Water began to withdraw from the collectors and flood the streets as a result of the unevenness.

On 16 July 1951, the city center was flooded for months due to the drainage system's inefficiency in removing rainwater. The water level reached two meters, and residents had to rely on boats to move around. The Hydrological Commission of the Valley of Mexico Basin recommended installing a system of pumps in the collectors to transport the water to the Grand Drainage Canal, a solution that would be costly due to the amount of electricity consumed.

The interceptor and west emitter were built in 1960 to receive and dislodge waters from the Nochistongo gorge. However, the city's rapid and steady growth rendered it insufficient. Lake Texcoco was 2 m below the city center in 1900, but 5.5 m above it in 1970 [36]. As a result, the federal government (Department of the Federal District) concluded, based on the findings of the General Directorate of Hydraulic Works investigations, that a drainage system was required that was not affected by subsidence and did not require the use of pumps, thereby lowering the cost of its operation. As the Tequixquiac tunnels were full, it was decided to build a fourth artificial outflow from the basin.

The first stage of the Deep Drainage System was built in 1967, during the presidency of Gustavo Díaz Ordaz, and was opened in 1975, with the government of Luis Echeverría. The first stage involved the construction of a 60 km tunnel with a 200 m depth and a discharge rate of up to 200 cubic meters per second. Luis Echeverría called this one of the century's most important works, one that would finally free the city from flooding [36].

An extension of 5.5 km to the central interceptor and a new interceptor of 16 km commenced during José López Portillo's six-year administration (1976–1982). The Deep Drainage System has been extended to more distant parts of the city, as has the Grand Drainage Canal, thus using existing infrastructure and linking the entire city. Under Carlos Salinas de Gortari (1988–1994), the Deep Drainage System covered 125 km.

For a long time, the drainage system was solely meant to remove rainfall and was closed annually for repair. In the mid-1990s, a combination system that dislodges both rainfall and sewage began to be deployed. Drainage work has persisted throughout the first two decades of the twenty-first century since the city continues to have flooding issues connected to expansion and subsidence.

The Deep Drainage System was expanded under Felipe Calderón Hinojosa's presidency. The East Emitter Tunnel (EET) was built in 2008 to increase system flexibility. In the municipality of Atotonilco de Tula in the state of Hidalgo, near the exit of the central emitter, the EET discharges up to 150 cubic meters per second. It is 62 km long, 7.5 m wide, and 55 to 150 m deep. A decade late and 20 million pesos over budget, it began operations in 2020. However, it is one of the world's largest drainage works.

In 2014, under Enrique Peña Nieto's administration, the first stage of the West II Emitter Tunnel was completed, reducing river overflows near Naucalpan, Tlalnepantla, Cuautitlán Izcalli, and Atizapán. It is 5573 km long, 12 to 110 m deep, and has a drainage capacity of 112 cubic meters per second. This project was nominated for Work of the Year in 2019 (see Figure 2).

Lake Texcoco went from covering almost the entire basin, about 7868 square kilometers at its origin [1], to 272.17 square kilometers in the mid-nineteenth century [40] and at present, the remnants of the lake cover an area of only 16.83 square kilometers within the federal area (see Figure 3).

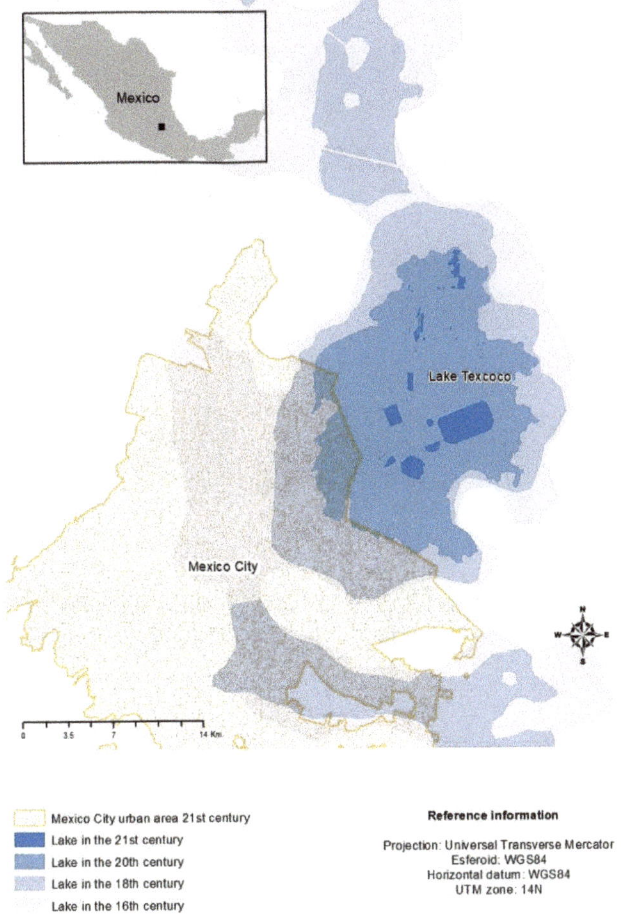

Figure 3. Historical evolution of the desiccation of Lake Texcoco. Own elaboration based on data from CONABIO [42] and Stangl [43], historical maps recovered from Mapoteca Orozco y Berra [44], and photo interpretation of satellite images from 2021.

4.5. Recent Climate Change in Mexico Basin

Mexico City's paradox is that it has built an artificial mega basin by importing water from the Lerma and Balsas basins to serve the city, while also expelling the waters through the constructed drainage system into the Tula basin (see Figure 4). Additionally, the extraction of water from the subsoil has resulted in the city sinking lower than the lake in respect to the difference from the place where it was located in pre-Hispanic times, leading to the burying of the city drainage deeper and deeper [45].

The delivery record of Cutzamala system (in the Balsas river basin) to Mexico Basin shows that, for the period 1997–2008, the average import was 15,162 cubic meters per second, while the corresponding record of the Lerma system shows that the respective import supply to the basin was 4231 cubic meters per second [46].

The ejected water is directed to irrigation district 003 in Tula, Hidalgo, which covers approximately 50,000 hectares and produces mostly corn (*Zea mays*) and alfalfa (*Medicago sativa*) [47] and, for the years 1989–1992, the yearly amount of exports to the Tula basin was 52,782 cubic meters per second [46]. It should be emphasized that Mexico City's and the metropolitan area's drainage system is of the combined sort (wastewater and rainwater), and that drainage expansion has continued in recent years, as has its water expulsion capacity.

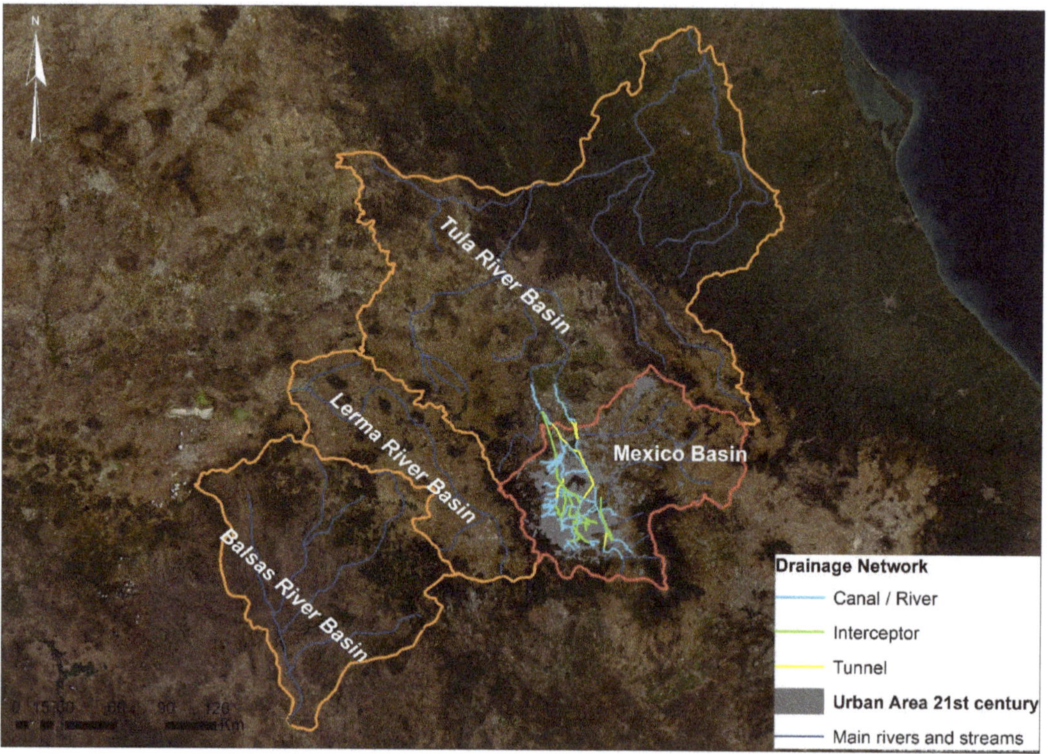

Figure 4. Megabasin created by alterations in the Basin of Mexico throughout history. Own elaboration based on data from CONABIO [42].

The Intergovernmental Panel on Climate Change (IPCC) anticipates that both environmental humidity and global temperature will increase [48]. Mexico City's transformation as a result of human-induced changes has resulted in an increase in the frequency and intensity of extreme events such as floods, droughts, and heat waves. According to Romero Lankao [49] between 1980 and 2006, there were 668 floods, 60 storms, 53 showers, 60 hail storms, 85 fires, and 14 droughts.

The National Water Commission's (CONAGUA) records on annual mean precipitation and annual mean temperature over the previous 37 years [50] indicate a decline in the amount of rainfall occurring in Mexico City annually and, in turn, an increase can be observed in the mean annual temperature (Figure 5).

Between 1920 and 1940, the Basin of Mexico had substantial population increase, particularly after 1940, and peripheral areas have been absorbed into the megalopolis. The forests and agricultural fields have vanished entirely, while there still exist portions of the lakes in the basin's south, they are endangered due to the pressures of urban water requirements.

According to Ezcurra et al. [45], numerous aquatic, subaquatic, and halophilic species have become extinct, while other types of vegetation are on the edge of extinction. Forest communities have been deeply affected in general; since 1985 over 9000 hectares of forest have been lost, and those that remain are threatened by pollution and pests. On the other hand, the introduction of exotic species to the basin's forests has resulted in the extinction of animal species that rely on indigenous plants for food and shelter. Eucalyptus trees, in particular, were planted extensively throughout the basin to increase evaporation and accelerate the lacustrine system's desiccation process.

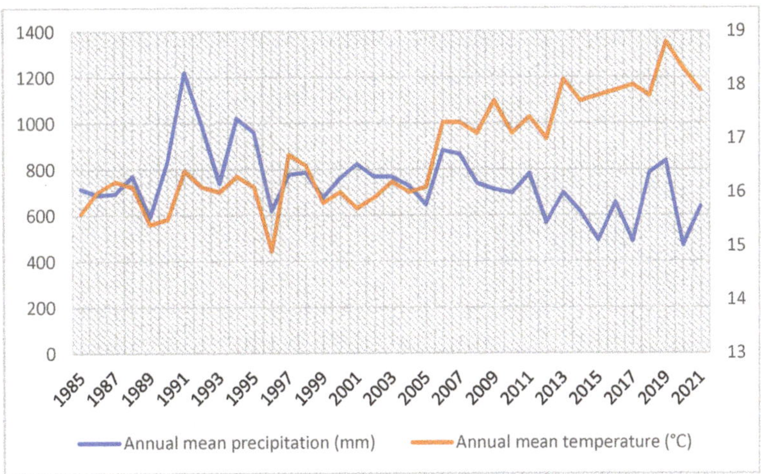

Figure 5. Graph of historical data on annual mean precipitation and annual mean temperature of Mexico City.

A study of changes in annual mean precipitation and temperature from 1950 to 2013 [51] indicated that average winter temperature and summer precipitation increased at an average rate of 0.1 degree Celsius per decade and 17.8 mm per decade, respectively. They forecast a temperature increase of between 1 and 3 degrees Celsius over the next 30 years, while rainfall is expected to increase and become more severe at the conclusion of the rainy season.

Clearly, the lake was abandoned in favor of urbanization. Following the Mexican Revolution, and particularly under Lázaro Cárdenas's (1934–1940) land reform, the desiccated territory surrounding Lake Texcoco was targeted for agriculture. They did not realize, however, that the fertility of the flooded land was attributable to the lake's presence.

Additionally, considerable efforts were made to restore what remains on the dry bottom of the former Lake Texcoco. CONAGUA initiated a program in the 1970s to plant halophytes on the lake's ancient bed in order to mitigate dust storms that afflicted the eastern part of Mexico's basin during the dry season. To retain a portion of the lake's ecological functions, a series of artificial lagoons with a combined surface area of 17 square kilometers were constructed. However, the government lost interest in this program, and for the last two decades, there has been a lot of debate about what to do with that area.

5. Discussion and Conclusions

For this study, segmenting the lake's history into periods marked by a critical moment in the country's social dimension enables the delineation of environmental policies that had a direct impact on Lake Texcoco, allowing for the tracking of changes in the use of this natural resource over time.

The indigenous Mexica were able to control the levels of the lakes by building dikes, roads, gates, and viaducts, which enabled them to live peacefully while taking advantage of the resources of the lake ecosystem through hunting, fishing, agriculture in chinampas, and transportation via the hydraulic system they had developed. While they utilized forests, altered and controlled the level of lakes, diverted rivers, and filled lakes in order to gain territory, their consequences were not as severe as those of the subsequent populations.

The pre-Hispanic era's key actors are the leaders of cities who assume the role of emperor; what stands out in Mexica society's urban design is the influence of their religious beliefs on their way of life, from the location of the city's foundation to the city's layout in relation to the aquatic environment, as Palerm [24] and Matos [2] indicate.

What occurred in Mexico during the Spanish colonization period is a reflection of the environmental approaches used in Europe at the time in countries such as Germany, Italy, France, and England [52], however few had lasting favorable effects.

The Spanish introduced new forms of agriculture, livestock, and forestry that had a negative impact on the natural landscape. The decision to build the capital of New Spain on the ruins of Tenochtitlan resulted in devastating floods and diseases among the local population, leading to the perception that the lakes were a danger.

Despite the fact that an attempt was made during the colonial period to use pre-Hispanic ways to manage the lakes, the transformation of the environment was so profound that they were insufficient, and the idea of drainage eventually won out over the other options. Enrico Martínez is best known for his design of the Huehuetoca tunnel and, later, the Nochistongo gorge, which resulted in the disruption of the basin's water balance and the creation of an outlet to the Tula River, respectively.

The tremendous floods that devastated the city at the colonial period prompted the viceregal government to make this choice, which resulted in the city being submerged under water for years. Making a snap decision in response to imminent natural events was a very risky move.

This analysis confirms Candiani [52], Vitz [23], and Tortolero's [21] assertion that colonization and the introduction of capitalism to Mexico resulted in the drying up of Lake Texcoco. This, combined with an unwillingness to live in a city surrounded by water, transformed the resource from a sanctuary to an unhealthy and dangerous environment for its residents.

As the nineteenth century progressed, the discussion over what to do with the lakes was still going on. During this period, three actors stand out who contributed to the speeding up of the drying process: the engineer Francisco de Garay, who made a number of interventions in the city's hydraulic system, the most notable of which was the design of the city's drainage system in 1856. This was authorized during Maximilian of Habsburg's empire in 1866 after being persuaded by Engineer Miguel Iglesias since this option would benefit, first and foremost, the interests of the group close to him.

Miranda Pacheco [38] and Tortolero [21] identify actors in this stage including doctors, engineers, specialist technicians, and members of Mexico City's top socioeconomic class during the Porfiriato era.

During the Porfiriato, the continuance of imperial works aimed to legitimize the state. It was a symbol of modernity and the power of the state that, far from providing a final solution to the city's drainage difficulties, was a determining element in the emergence of new environmental problems and the hunt for more technical solutions to floods in the years after its construction.

Deep drainage was required as a result of the collapse of the Porfirian drainage system, which has continued to spread to this day. History appears to be repeating itself, but in ever bigger and even more tragic proportions in light of the rapid population development and ecological deterioration of the basin in recent decades.

In modern Mexico, the state and businesspeople emerge as actors. The inclination of political actors, particularly in the post-revolutionary period, to exploit Lake Texcoco to construct a new political–social order is noticed; with Cárdenas, the use of natural resources developed in tandem with social change, what Boyer and Wakild [20] refer to as "Social Landscaping".

When Lake Texcoco dried up, it exposed the lands at its bottom, allowing the urban area to expand. However, in recent years, the remnants of Lake Texcoco have been put on the table for discussion over their future uses, which has sparked opposition from some members of the community. However, before making any further changes to the lake's ecosystem, it is critical to consider the cycle of historical decisions that have resulted in one environmental calamity after another.

Blaikie and Brookfield's study [9] is a seminal work in political ecology for explaining land deterioration. Using this as a starting point, a model was built to depict the cumulative decisions on Lake Texcoco that resulted in environmental degradation. The accumulation of hydraulic projects necessary to "keep the city safe" has degraded the ecosystem. Political economy is critical, since it controls the dynamics of changes in the country's development, which is directly mirrored in Mexico City and its environs.

For example, in post-revolutionary modern times, when floods continued to occur despite the completion of the big drainage canal, the choice to continue developing Mexico City's drainage system and expelling all water from the city, whether rainfall or sewage, prevailed.

Mexico City has experienced flooding continuously, and while it once possessed vast water resources, it now faces a high vulnerability to climate change and the inability to meet the city's water demand. The examination of decisions made on Lake Texcoco and the identification of individuals enables this to be attributed to environmental changes and changes in the hydrological cycle.

The change in the hydraulic cycle has been profound and irreversibly altered the ecosystem, resulting in a paradoxical situation in which floods continue despite the sophistication of the drainage system, while nearly a third of drinking water must be imported from increasingly greater distances. Thus, in addition to impacting the basin's climate, the alteration of the water system has impacted the regional water balance and availability.

Throughout the basin's environmental transition, misinformed decisions, combined with political, economic, and social influences, have had a detrimental impact on natural systems. Mexico City, like cities such as Dhaka, Beijing, and Sao Paulo, relies heavily on groundwater and interbasin water transfers for water supply [53]. Drying, drainage, and transfer projects are excellent examples of how social priorities functioned and evolved over time. These cities are particularly revealing about how humans interact with nature and affect ecology, land use, and water resources.

Author Contributions: Conceptualization, C.M.-R., E.O.-T. and V.E.-H.; data curation, C.M.-R. and E.O.-T.; investigation, C.M.-R.; methodology, C.M.-R. and E.O.-T.; Supervision, V.E.-H.; validation, D.F.-R., M.C.D. and G.S.B.V.; visualization, C.M.-R.; writing—original draft, C.M.-R.; writing—review and editing, C.M.-R. and E.O.-T. All authors have read and agreed to the published version of the manuscript.

Funding: This research received no external funding.

Institutional Review Board Statement: Not applicable.

Informed Consent Statement: Not applicable.

Data Availability Statement: Publicly available datasets were analyzed in this study. Geographic information can be found here: CONABIO. 2022. National Biodiversity Information System. http://www.conabio.gob.mx/informacion/gis/ (accessed on 14 March 2022). Climatological information can be found here: CONAGUA. 2022. Monthly Temperature and Rainfall Summaries: https://smn.conagua.gob.mx/es/climatologia/temperaturas-y-lluvias/resumenes-mensuales-de-temperaturas-y-lluvias (accessed on 14 March 2022). Stangl, W., & (FWF), A. S. F. (2019). Data: Lago Texcoco, 18th century (V1 ed.). Harvard Dataverse. https://doi.org/doi/10.7910/DVN/DOT5EY (accessed on 14 March 2022).

Acknowledgments: The authors thank the National Science and Technology Council (CONACYT) for the scholarship awarded for the doctoral studies from which this research is derived.

Conflicts of Interest: The authors declare no conflict of interest.

References

1. Alcocer, J.; Williams, W.D. Historical and Recent Changes in Lake Texcoco, a Saline Lake in Mexico. *Int. J. Salt Lake Res.* **1996**, 5, 45–61. [CrossRef]
2. Matos Moctezuma, E. *Tenochtitlán*, 3rd ed.; Fondo de Cultura Económica: Mexico City, Mexico, 2016; ISBN 9681681185.
3. de Garay, F. *El Valle de México: Apuntes Históricos Sobre Su Hidrografía, Desde Los Tiempos Más Remotos Hasta Nuestros Días*, 1st ed.; Secretaría de Fomento: Mexico City, Mexico, 1888.
4. Secretaría de Fomento Memoria Histórica, Técnica y Administrativa de Las Obras Del Desagüe Del Valle de México, 1449–1900. In *Junta Directiva del Desagüe del Valle de México*, 1st ed.; Tipografía de la Oficina Impresora de Estampillas: Mexico City, Mexico, 1902; Volume 1.
5. Offen, K. Political Ecology History. *Hist. Geogr.* **2004**, 32, 19–42.
6. Khan, M.T. Theoretical Frameworks in Political Ecology and Participatory Nature/Forest Conservation: The Necessity for a Heterodox Approach and the Critical Moment. *J. Political Ecol.* **2013**, 20, 460–472. [CrossRef]
7. Bailey, S.; Bryant, R. *Third World Political Ecology: An Introduction*; Routledge: London, UK, 2005; ISBN 0203974360.
8. Blaikie, P. Changing Environments or Changing Views? A Political Ecology for Developing Countries. *Geography* **1995**, 9, 203–214.
9. Blaikie, P.; Brookfield, H. *Land Degradation and Society*; Routledge: London, UK, 2015; ISBN 1315685361.
10. Neumann, R.P. Political Ecology: Theorizing Scale. *Prog. Hum. Geogr.* **2009**, 33, 398–406. [CrossRef]
11. Mathevet, R.; Peluso, N.L.; Couespel, A.; Robbins, P. Using Historical Political Ecology to Understand the Present: Water, Reeds, and Biodiversity in the Camargue Biosphere Reserve, Southern France. *Ecol. Soc.* **2015**, 20, 1–13. [CrossRef]
12. Robbins, P. *Political Ecology: A Critical Introduction*; John Wiley & Sons: Hoboken, NJ, USA, 2019; ISBN 1119167442.
13. Ojeda Trejo, E. GIS and Land Use in Texcoco Municipality, Mexico: Contrasting Local and Official Understandings. Ph.D. Thesis, Durham University, Durham, UK, 2001.
14. McNeill, J.R. Observations on the Nature and Culture of Environmental. *Hist. Theory* **2003**, 42, 5–43. [CrossRef]
15. Gutiérrez Castorena, M.; Ortiz Solorio, C. Origen y Evolución de Los Suelos Del Ex Lago de Texcoco, México. *Agrociencia* **1999**, 33, 199–208.
16. Aguirre Gómez, R. *Estudios Sobre Los Remanentes de Cuerpos de Agua En La Cuenca de México*; UNAM, Instituto de Geografía: Ciudad de Mexico, Mexico, 2010; ISBN 9786070219887.
17. Candiani, V. The Desagüe Reconsidered: Environmental Dimensions of Class Conflict in Colonial Mexico. *Hisp. Am. Hist. Rev.* **2012**, 92, 5–39. [CrossRef]
18. von Humboldt, A. *Ensayo Polítco Sobre La Nueva España*, 1st ed.; Jules Renuard: Paris, France, 1827.
19. Lattes, E.L. Historia Del Desagüe Del Valle de México. *Ing. Hidráulica En México* **1988**, 3, 66–68.
20. Boyer, C.R.; Wakild, E. Social Landscaping in the Forests of Mexico: An Environmental Interpretation of Cardenismo, 1934–1940. *Hisp. Am. Hist. Rev.* **2012**, 92, 73–106. [CrossRef]
21. Tortolero, A. El Agua y La Historia Medioambiental: Revisión Historiografica y Estudios de Caso. *Iztapalapa* **2001**, 50, 425–454.
22. Iracheta, P.; Dávalos, M. La Historia Del Agua En Los Valles de México y Toluca. *Historias* **2004**, 57, 109–130.
23. Vitz, M. *A City on a Lake. Urban Political Ecology and the Growth of Mexico City*; Duke University Press: Durham, NC, USA, 2018; ISBN 9780822372097.
24. Palerm, Á. *Obras Hidráulicas Prehispánicas En El Sistema Lacustre Del Valle de México*, 1st ed.; INAH: Mexico City, Mexico, 1973.
25. Matos Moctezuma, E. Posclásico Tardío (1350–1519 d.C.): El Dominio Mexica. *Arqueol. Mex.* **2007**, 15, 58–63.
26. Mann, C. *1491: New Revelations of the Americas before Columbus*, 2nd ed.; Vintage: New York, NY, USA, 2006; ISBN 1400032059.
27. Neumann, R.P. Political Ecology. In *International Encyclopedia of Human Geography*; Kitchin, R., Thrift, N., Eds.; Elsevier: Oxford, UK, 2009; pp. 228–233, ISBN 978-0-08-044910-4.
28. Biersack, A. Reimagining Political Ecology: Culture/Power/History/Nature. In *Reimagining Political Ecology*; Duke University Press: Durham, NC, USA, 2006; pp. 3–40.
29. Zimmerer, K.S. Methods and Environmental Science in Political Ecology. In *The Routledge Handbook of Political Ecology*; Routledge: London, UK, 2015; pp. 172–190, ISBN 1315759284.
30. Khan, M.T. The Nishorgo Support Project, the Lawachara National Park, and the Chevron Seismic Survey: Forest Conservation or Energy Procurement in Bangladesh? *J. Political Ecol.* **2010**, 17, 68. [CrossRef]
31. Carballal Staedtler, M.; Flores Hernández, M. Elementos Hidráulicos En El Lago de México-Texcoco En El Posclásico. *Arqueol. Mex.* **2004**, 12, 28–33.
32. Marroquí, J.M. *La Ciudad de México*, 1st ed.; Tipografía y Litografía "La Europea": Mexico City, Mexico, 1900.
33. García Martínez, B. Conquista (Siglo XVI, a Partir de 1519): Cambios y Continuidades. *Arqueol. Mex.* **2007**, 15, 64–68.
34. Espinosa Castillo, M. *Ecatepec y Nezahualcóyotl de Suelos Salitrosos a Ciudades de Progreso*, 1st ed.; Secretaría de Educación del Estado de México: Mexico City, Mexico, 2010; ISBN 978-607-495-033-5.
35. Valdés Valek, G. *Agua: Reflejo de Un Valle En El Tiempo*; UNAM: Mexico City, Mexico, 2000.
36. Cohen, P.M.; González Reynoso, A. *¿Guerra Por El Agua En El Valle de México? Estudio Sobre Las Relaciones Hidráulicas Entre El Distrito Federal y El Estado de México*, 1st ed.; UNAM, Coordinación de Humanidades, PUEC, Fundación Friedrich Eber: Mexico City, Mexico, 2009; ISBN 970-32-2968-9.

37. Mora, J.M. *Memoria Para Informar Sobre El Origen y Estado Actual de Las Obras Emprendidas Para El Desagüe de Las Lagunas Del Valle de México, Presentó a La Exma. Diputación Provincial El Vocal Dr. D. José María Mora, Comisionado Para Conocerlas*; Imprenta de la Aguililla: Mexico City, Mexico, 1823; ISBN 9789896540821.

38. Miranda Pacheco, S. Desagüe, Ambiente y Urbanización de La Ciudad de México En El Siglo XIX. *Relac. Estud. Hist. Soc.* **2019**, *40*, 31–72. [CrossRef]

39. Espinosa, L. Informe Sobre Las Modificaciones de La Sección Del Túnel de Tequisquiac y de Su Pendiente. In *Memoria Técnica y Administrativa de las Obras del Desagüe del Valle de México*; de Fomento, S., Ed.; Tipografía de la Oficina Impresora de Estampillas: Mexico City, Mexico, 1879; pp. 104–118.

40. Legorreta, J. *El Agua y La Ciudad de México. De Tenochtitlán a La Megalópolis Del Siglo XIX*, 1st ed.; Universidad Autónoma Metropolitana: Mexico City, Mexico, 2006; ISBN 970310522X.

41. Reyes, A. *Visión de Anahuac*, 1st ed.; Fondo Editorial de Nuevo León: Monterrey, Mexico, 2009.

42. Geoportal Del Sistema Nacional de Información Sobre Biodiversidad [15,793]—CONABIO. Available online: http://www.conabio.gob.mx/informacion/gis/ (accessed on 19 March 2022).

43. Stangl, Werner, 2019, Data: Lago Texcoco, 18th century, Harvard Dataverse, V1. Available online: https://doi.org/10.7910/DVN/DOT5EY (accessed on 7 March 2022).

44. Mapoteca Manuel Orozco y Berra. Available online: https://mapoteca.siap.gob.mx/ (accessed on 19 March 2022).

45. Ezcurra, E.; Mazari-Hiriart, M.; Pisanty, I.; Aguilar, A.G. *The Basin of Mexico: Critical Environmental Issues and Sustainability*; United Nations University Press: New York, NY, USA, 1999.

46. Gómez-Reyes, E. Valoración de Las Componentes Del Balance Hídrico Usando Información Estadística y Geográfica: La Cuenca Del Valle de México. *Rev. Int. Estadística Geogr.* **2013**, *4*, 5–27.

47. Oviedo, F.M.C.; Herrera, M.L.; Hernández, R.B.; Sandoval, O.A.A.; Lucho-Constantino, C.A.; Santamaría, M.I.R. Degradación Del Suelo En El Distrito de Riego 003 Tula, Valle Del Mezquital, Hidalgo, México. *Rev. Científica UDO Agrícola* **2012**, *12*, 873–880.

48. IPCC. *Climate Change 2021: The Physical Science Basis. Contribution of Working Group I to the Sixth Assessment Report of the Intergovernmental Panel on Climate Change*; Masson-Delmotte, V., Zhai, P., Pirani, A., Connors, S., Péan, C., Berger, S., Caud, N., Chen, Y., Goldfarb, L., Gomis, M., et al., Eds.; Cambridge University Press: Cambridge, UK; New York, NY, USA, 2021.

49. Romero Lankao, P. Water in Mexico City: What Will Climate Change Bring to Its History of Water-Related Hazards and Vulnerabilities? *Environ. Urban.* **2010**, *22*, 157–178. [CrossRef]

50. CONAGUA Resúmenes Mensuales de Temperaturas y Lluvia. Available online: https://smn.conagua.gob.mx/es/climatologia/temperaturas-y-lluvias/resumenes-mensuales-de-temperaturas-y-lluvias (accessed on 7 March 2022).

51. Behzadi, F.; Wasti, A.; Haque Rahat, S.; Tracy, J.N.; Ray, P.A. Analysis of the Climate Change Signal in Mexico City given Disagreeing Data Sources and Scattered Projections. *J. Hydrol. Reg. Stud.* **2020**, *27*, 100662. [CrossRef]

52. Candiani, V.A. *Dreaming of Dry Land: Environmental Transformation in Colonial Mexico City*; Stanford University Press: Stanford, CA, USA, 2014.

53. Li, E.; Endter-Wada, J.; Li, S. Characterizing and Contextualizing the Water Challenges of Megacities. *J. Am. Water Resour. Assoc.* **2015**, *51*, 589–613. [CrossRef]

MDPI

St. Alban-Anlage 66

4052 Basel

Switzerland

Tel. +41 61 683 77 34

Fax +41 61 302 89 18

www.mdpi.com

Land Editorial Office

E-mail: land@mdpi.com

www.mdpi.com/journal/land

www.ingramcontent.com/pod-product-compliance
Lightning Source LLC
LaVergne TN
LVHW070919170726
843515LV00005B/824